Keep Your Personal Computer Running Like New

Avram Grossman

A PC Users Guide for Avoiding the Grief of Losing Your Information

Published by E-Novative, Inc.
Avram S. Grossman

First Printing, June 2020

Printed in the United States of America

ISBN 978-1-7351123-0-5

This ideas and suggestions presented in this work are from the author alone. It is the objective of this work to educate and inform the reader so that they may learn and understand what goes into the technology that makes their personal computer work. It does not necessarily imply or make any assurances that the reader will be able to maintain their personal computer trouble-free. Sometimes things just happen.

Contents

Table of Figures

Introduction

Would you like to learn how you can keep your computer running smoothly and efficiently?

Would you like to learn how to avoid making mistakes that could lead to a slow, lethargic computer, or worse—lost information and data?

Would you like to understand what is inside your computer and how it works?

Do the lights, buttons, cables, and software choices intimidate you?

This book is here to help you understand each component of the computer and the functions they perform.

You will learn respect for your personal computer.

You will become a computer guru and impress your family with your new knowledge. You'll know why computers misbehave and learn some simple steps to avoid problems and keep your personal computer (PC) working like new.

Personal computers are an essential tool in your daily life!

My IT career has centered on design and implementation of new technology. I have designed computer and electronics hardware, starting with very large main-frame computers. I have worked on minicomputer architecture and designed micro-computer-based hardware, operating systems, and application software. As an IT Consultant for over twenty years, I have seen just about everything that can go wrong with a computer. More importantly, I have seen just about everything a person can do to their computer to provoke a real disaster! I might add that just when I'd thought I'd seen it all, the next client presents a problem stranger than the last.

Over the past twenty years, I have replaced my desktop computer just a few times. It is typical for me to use the same PC for five or six years. I did have to replace my laptop once, but only after it was stolen out of my car. I guess that doesn't count. On a rare occasion something did fail on its own, but in most cases I was able to repair the problem and keep the computer and software running for many more years. For example, a power supply died, and once I experienced a hard drive crash. A replacement power supply cost $45 and took about twenty minutes to replace, and the hard drive was configured as RAID-0 (two drives that mirror one another) so no data was lost, as the second drive contained a mirror of the first.

I have installed many new computers for my clients. I don't know why, but it seems when an employee of a company uses a computer that they don't own, they appear to treat their computer with carelessness. Is it that they don't have the basic knowledge and experience to work kindly with their office computer? Or maybe they just don't take the few extra minutes to do things the preferred way? I can often tell which client's employees will be calling me within a few months to say their computer is acting strangely. In almost all situations this strange behavior is what we call in the industry user error. They are trying to do things with their computer that it wasn't meant to do. Kind of like trying to put your feet into new shoes and then taking them back to the store because you think they are defective—or, it might be that you need to put the left foot into the shoe designated for your left foot, and same for the right shoe, for them to operate correctly.

At a recent business professional group in Newport Beach I met an information technology (IT) consultant. He introduced himself to the group and asked the audience, "Do you own a computer?" Everyone responded, "Yes". He then said, "You need to call me!" Why did this professional tell people they need to call him if they own a PC?

Both he and I agree that ninety percent of the calls we get from our IT clients is to fix things that have gone wrong with their personal computer, and most of these problems were probably avoidable. Another colleague said to me, "Computers don't break themselves. We do it."

I depend on my computer to be ready to work for me whenever I need it. I don't have the time or patience to have to fiddle with it each time I want to use it. I have work to get done and often limited time to do it in. I don't want to take the time to reinstall all the application programs and data and set up my personal configurations and preferences on a new computer. I'd rather keep my little workhorse happy and running smoothly. If you purchase a new computer without a basic understanding of how to work with it, having a new machine will not prevent the same problems from reoccurring as they did with your last computer.

IT professionals whose job it is to service other people's computers usually have personal computers that seem to work forever, because they have the knowledge and experience in how to do things the correct way and troubleshoot problems when they arise. The goal, therefore, is to teach you the skills you need to troubleshoot basic problems yourself.

Why do IT professionals have computers that seem to run forever, and you keep having problems with yours? The answer lies in this simple adage:

If you take care of your technology, your technology will take care of you.

You can apply this adage to anything mechanical or technical in your life: your car, your bicycle, your cell phone, the vacuum cleaner, most home appliances, the lawn mower and yes, even your personal computer. If you take care of your things, they will continue to take care of you.

The main reason I wrote this book was so that you could learn how to avoid most problems and how to diagnose and fix problems when they arise. By learning how to use your computer with kindness and care, you should be able to avoid having needless problems that run up the cost for getting help. It might save you the grief of having your PC stop working all together and losing access to the valuable information stored on the hard drive.

Keeping your computer running trouble free forever is a great goal unto itself, but there is another more important objective. The reason you might want to keep your computer running forever is because, like many of us, your entire life's information and data is stored in the computer. We use personal computers for just about everything. I replace my computers every five to six years, because I want and need the newest processors and larger disk drive storage capacity for my software development work.

We store family photos for sharing on our computers. We use our computers to send and receive email to all our family members, friends, and for business transactions. From time to time we go back and read what was said. We use computers to access our personal and business banking websites. Those website addresses and often the associated usernames and passwords are retained, or cached, in the browser—which we would otherwise forget. We use computers for word processing to compose letters we've written and sent. We use computers to write and save letters never meaning to send them (lyrics from the Moody Blues.) We use calendar and scheduling software to reminder us of birthdays, anniversaries, and upcoming events. We use Skype, Zoom, and other social media services to keep in communication with our family, friends, and yoga groups. You get the idea—for just about everything you do, you use a personal computer to make your life easier.

The list of things we use our personal computer for is endless. Just when you think you've found all the things you can use a PC for, someone comes up with a new application for some new purpose.

You spend countless hours of time in front of the monitor and keyboard entering, reading, organizing, and searching for information. You would go crazy if your computer stopped working. You might go even crazier if you lose the information you thought you had saved someplace, only to discover the backup wasn't backing up correctly.

These are some of the reasons why you want to keep your computer running forever, so when you sit down at the keyboard and need to get some work done, your computer is ready to work with you, not against you. When you are ready to migrate up to a better and faster computer, all your information will migrate to the new computer smoothly. You can replace your old computer while it is still working, and when you are ready to do so, not when you are forced to get a new one due to a crash.

Finally, one more thing before you get started: please don't be afraid of your computer. There are very few mistakes you can make that would be fatal and that could not be undone. It's okay to make changes to settings in the operating system. In many situations you will want to change the default setting anyway to make improvements in how you use your PC and how it works best for you. Remember, a PC is your *personal* computer, so personalize it.

As you learn to trust yourself using the computer, your computer will gain confidence in your skills to use it. You read that correctly. It sometimes seems that our computer knows we are in a rush and not feeling confident, and it responds in kind. As you apply what you learn in this book, facets of the computer you find frustrating or difficult to understand will begin to appear easy. In no time your friends will be calling on you for advice.

As you read through the chapters, I will make suggestions for some preferred methods or procedures. The Suggestions are also listed in the Appendix. Look for the Suggestion Icon

Anatomy of a Personal Computer

It is my personal belief that the more you know how and why things work, your understanding and appreciation of the things around you will grow and get easier.

I apply this philosophy to just about everything I encounter, not just computers. Ever since I was knee high to a soldering iron I was taking things apart. My curiosity would lead me to ask questions like: How was this made? How does it work? Why does it work? It worked before, why is it not working now? Can it be fixed?"

This curiosity also led me to start taking things apart just to see if I could put them back together. When I buy some new technical gadget, I want to see what all the features are inside the device. I explore every menu option and change just about every setting, just to see what it does and how it changes the device. Once, after having turned the news announcer green on my new smart TV, I found the *reset all* software button in one of the menus. Everything magically went back to default settings and began working normally again. You can thank your lucky stars there are *reset all* and default-settings buttons. Knowing what the various components are, what function they provide, and how they interact with the other parts is helpful to personalize the device. It helps you to diagnose and resolve issues and customize your computer for your specific preferences.

WHAT IS THAT CALLED?

A former marketing director at Computer Automation, Inc., (1982), Bill Pitts, used to refer to every component in our minicomputer line as a "framistan." As a marketing person, he really wasn't concerned with knowing what each part was actually called or what function it performed. He was great at sales calls, explaining to potential customers that our minicomputers featured the latest version of framistan and that our competition did not offer it.

When you call a repair technician to come fix something you know little about, they might as well be telling you the framistan needs replacing. He might also tell you "Oh, they are also very expensive to replace." So before you call in someone to service your PC, it might be helpful to know what to tell them you suspect is wrong. Otherwise a more unscrupulous repair technician might tell you what they want to fix or replace for you and charge you accordingly, without concern for what is actually wrong.

As an example, a local cable company that services a large senior retirement community was telling their senior citizen customers that the modems were running slowly because they have become obsolete and needed to be replaced. With no understanding of how a cable modem worked, most of the customers said, "Sure, I'll buy the new modem." Those folks who contacted me through a word-of-mouth referral asked me to look at their modems to confirm if they really needed a new cable modem. New cable modems were being sold for over $120.00. For those on a fixed income, this was too much to spend. I was able to upgrade the modem firmware, which was available as a free download from the vendor's website. The upgraded firmware improved the performance of these devices that the company had called obsolete. I also didn't charge for my help. If you don't know what is wrong and what you might do to fix the problem, be prepared to pay for something you don't really need.

Many people know nothing more about their computer than to call it a computer, a laptop, a tablet, or an iPad. When a client calls to let me know their computer isn't working, it sometimes takes a bit of effort and patience to get them to explain to me which component they are talking about that isn't working. Having knowledge of the names of the parts of the computer, and the terminology to describe menus, icons, and buttons on the screen is helpful. Being told "the green thing in the corner is gone" is not useful.

"Oh, you mean the cursor on the screen isn't moving when you move the mouse?" Or "You don't see anything on the monitor at all? Is the power turned on?" Yes, I've actually had clients call me because they didn't remember they turned off their monitor the night before and this morning their screen was blank.

This chapter is written so you'll know what the various parts and components are called and what function they serve. You will be able to speak knowledgably about what those parts do and how they interact with one another, and what component might be attributed to something not working correctly. If something isn't working as you expect, you will learn to avoid the initial panic attack and use your knowledge to compile a list of possible reasons and solutions. For example, if your computer seems to be running slowly, you will understand it's not because you have too many images on the hard drive. You'll instead suspect it might be a virus or Trojan parasitic software that is eating up your central processing unit (CPU) cycles, and you'll know how to diagnose and resolve the issue before deleting all your photos in vain. When something appears wrong, it's not the end of the world as you know it, if you know how to diagnose and correct the issue.

HARDWARE

Let's start by looking at what parts the computer has inside the cover. All PCs are made up of the same basic components or building blocks. It doesn't matter if you have a Windows PC or a Mac, an android mobile device, or iPad. They are all built upon similar building blocks. These building blocks of those physical parts and components are affectionately referred to as the hardware.

Hardware, like the name suggests, is comprised of the physical pieces of sheet metal and plastic and are put together with screws, rivets, and snap-together parts. Hardware houses the electronics which consist of the microprocessor, various electronic circuits with associated resistors and capacitors, memory chips and storage devices, input/output controllers, and a power supply with cables and connectors that provide voltage to everything.

When we speak of the hardware, this includes the chassis or enclosure, and all the stuff inside the enclosure. Let's take a look at what the hardware consists of:

The Chassis

This is the enclosure or box that holds everything in one package. The form factor is a term used to describe the physical shape and size of the enclosure. Computer form factors come in a mini tower and midsize tower form factor (see Figure 1) and could apply to a laptop, PC, and tablet form factors. With the advent of super scale integration, entire PCs are available as small as 4"W x 4"L x 0.75"H, require very low power, and don't require cooling fans (see Figure 2: Fanless Chromebox). As shown in Figure 3, I use 19" wide rack-mounted computers as web servers in the data center which are 2" tall and 24" deep.

The chassis encloses all the components that make up the computer system. Now let's take a look inside the chassis or enclosure.

The Power Supply

Electronic components in a PC need regulated power to operate. All PC power supplies today use a technology called a switching power supply to provide regulation of each of the required voltages and current levels needed by each of the parts. The switching power supply components are all built into a self-contained metal housing that is mounted inside the chassis as shown in Figure 4. A standard three-pronged A/C power connector is located on the back of the power

Figure 1: Mini-Tower PC

Figure 2: Fanless Chromebox

Figure 3: Rack-Mounted Server

supply that is plugged into the wall outlet or battery backup device. This is the power input.

The various output voltages are distributed to a few places on the motherboard and to each of the disk drives and peripheral devices inside the chassis. The shielded power cables resemble Medusa's snakes. Different types and sizes of connectors at the end of each cable deliver the voltages needed for that type of device. A typical switching power supply delivers multiple voltages of +3.3VDC, -12VDC, +12VDC, and +5VDC. The power supply has a built-in fan to help keep airflow moving over its components to keep it running cool. The fan is also a source of dust getting into the power supply and chassis, and could become a source of noise.

The power supply's fan is prone to wear and can eventually cause an annoying squealing or clanging sound.

Figure 4: Switching Powering Supply

Power supplies for tower and desktop chassis are mostly standardized. They all have the same type of connectors, although the higher-wattage power supplies used in gaming or high-end graphics computers might have more power connectors than a standard PC power supply. Should a power supply fail, they cost under $60 and are easy to replace. If a power supply does fail, don't throw out the computer. Replace the power supply.

There are very small computers on the market that are very low power and don't require a switching power supply. Instead, they are powered by a single voltage power (+5V) adapter, much like the ones that are used to charge mobile phones. These fanless mini PCs are great for use as a streaming media box, such as a Chromebox. I use a fanless Windows 10 mini PC connected via an HDMI cable to a widescreen television for watching Netflix and Amazon Prime.

The Motherboard

This is the main component in the enclosure that comprises the computer. It's actually not a component, but rather an assembly of the most crucial electronic components that perform the various necessary functions. As seen in Figure 5, the motherboard contains the microprocessor chip or CPU, memory chips, disk controller chip, input/output (I/O) controller chips, power regulator chips, and the connectors for the peripheral devices. All of these electronic chip components and connectors are surface mounted on to a multilayered highly dense printed circuit board.

Figure 5: Typical PC Motherboard

In the early days of the personal computer, the components that went into building the motherboard were pretty simple as compared to today's highly integrated chip technology. The functional components like the memory, monitor display, keyboard interface, and hard disk controllers were often built on separate cards that plugged in to slots. When you bought a PC, you selected the basic CPU speed and form factor, and then you could pick and choose the size of memory and external storage or peripherals (floppy disk and hard disk devices), and you could easily upgrade your PC as new cards became available. Today, these functional units are highly integrated circuits designed into one chip and mounted onto the motherboard; they are generally not upgradable.

What's on a Motherboard?

The motherboard is a complex printed circuit board that contains all the semiconductor components. Each of the components is designed using super-scale integration technology. That is, they pack a lot of functionality in a single chip, with several chips required for specific tasks. Refer to Figure 6 for a deep look into the block diagram showing what the motherboard contains.

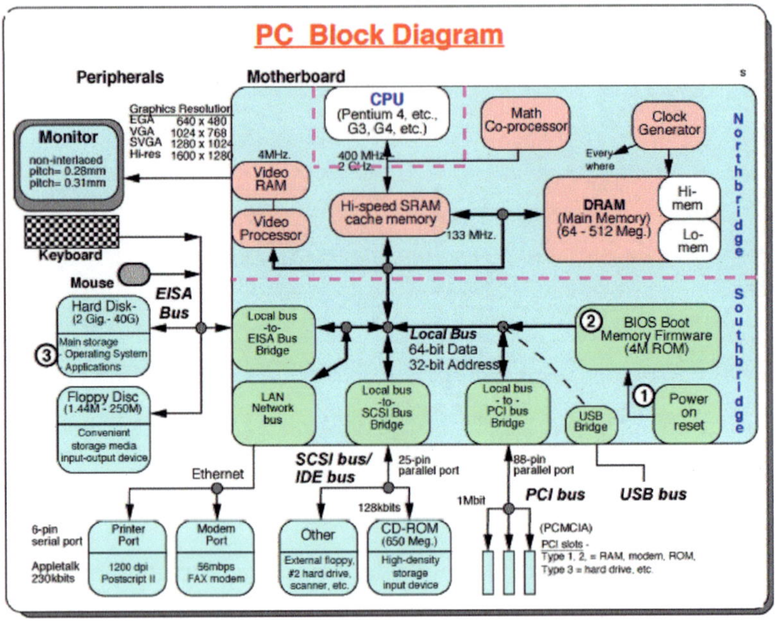

Figure 6: Basic Block Diagram of a PC Motherboard

Let's take a tour of the motherboard and explore what we'll find there:

The Microprocessor: This is called the central processing unit (CPU). It has been affectionately called the brains of the computer. The CPU is the workhorse of the computer and is where all the processing is done.

A CPU core is the CPU's processor. In the old days, every processor had just one core that could focus on one task at a time. Today, CPUs have been two and eighteen cores, each of which can work on different tasks at the same time.

A single core can work on one task while another core works a different task, so the more cores a CPU has the more efficient it is. Many processors, especially those in laptops, have two cores, but some laptop CPUs (known as mobile CPUs), such as Intel's 8th Generation processors, have four cores.

Processors use a method called simultaneous multithreading, or if it's an Intel processor they call it Hyper-Threading Technology, to split a core into virtual cores, which are called threads. For example, AMD CPUs with four cores use simultaneous multithreading to provide eight threads, and most Intel CPUs with two cores use Hyper-Threading to provide four threads. This means the microprocessor is capable of an increase in processing ability without an increase in power consumption.

Some application programs take better advantage of multiple threads than others. Lightly-threaded apps, like computer games, don't benefit a lot from multi-core processors. Most video editing and animation programs do take advantage of multi-core processors can run much faster with extra threads. That is one reason why gaming computers cost more.

Figure 7: Intel CPU Processor Chip

Depending upon the type of CPU chip on the motherboard, the CPU chip can consume a significant amount of power and therefore generate a significant amount of heat. Most microprocessors have a heat-sink mounted on top of the chip to dissipate heat, and many heat-sinks also have a small CPU fan mounted on it. Refer to Figure 8.

Figure 8: CPU with heatsink and cooling fan

Memory: Memory is where the software and data resides when the CPU is running and processing information. When power to the computer is off, the contents of the memory is empty. You'll learn how and when the memory is initialized in the ROM BIOS startup section on Page 28. The memory chips are actually not on the motherboard. Rather, the memory cards are discrete assemblies as shown in Figure 10 and are installed on the motherboard in memory connector slots as shown in Figure 9.

Figure 9: PC Memory Location on Motherboard

Memory comes in various types. The most popular is "RAM," which stands for random access memory. That means that any piece of data stored in the memory can be randomly addressed or stored and retrieved. In the early days of memory, the memory storage devices only allowed data to be written and read sequentially, called first-in, first-out memory. They were very slow. I mention this only to show you how memory technology has evolved in just a few short years.

Error-correcting code (ECC) memory is now the standard type of memory storage chip. It is specifically designed to detect, correct, and self-monitor most common kinds of internal data corruption. As data is processed, ECC memory equipped with a special algorithm constantly scans and corrects single-bit memory errors. This ensures that no erroneous or corrupt data is accidentally stored in memory.

The RAM chips are mounted on printed circuit cards that are then snapped into vertically oriented connectors in rows on the motherboard. Depending on the make and model of your PC, there will be a number of slots so you can change out, upgrade or replace memory modules.

Figure 10: Typical Add-in Memory

RAM comes in various technologies, including static RAM (SRAM), dynamic RAM (DRAM), synchronous dynamic RAM (SDRAM), single data rate synchronous dynamic RAM (SDR SDRAM), double data rate synchronous dynamic RAM (DDR SDRAM, DDR2, DDR3 and DDR4), graphics double data rate synchronous dynamic RAM (GDDR SDRAM, GDDR2, GDDR3, GDDR4, GDDR5) and flash memory. You might be tested on this later.

What's in a byte?

Memory chips come in various capacities that are measured in gigabytes. A single byte is a sequence of eight bits, enough to represent one alphanumeric character, processed as a single unit of information. A single letter or character would use one byte of memory (eight bits) and two characters would use two bytes (sixteen bits).

In computer speak the bits are represented as "on" or "1" and "off" or "0". There are eight bits in a byte, and a byte is used to pass data in its basic form—a character.

An alphanumeric character (for example a letter or number such as A, B, or 7) is stored as one byte. To store the letter R the computer uses the eight bits as 01010010.

A document containing one hundred characters would use one hundred bytes (eight hundred bits)—assuming the file didn't have any overhead (additional data about the file which forms part of the file). Many non-alphanumeric characters such as special symbols and foreign language characters use multiple bytes.

How many bytes is that? The nomenclature for data sizes is as follows:

1,024 bytes	=	1 KB (kilobyte)
1,024 KB	=	1 MB (megabyte)
1,024 MB	=	1 GB (gigabyte)
1,024 GB	=	1 TB (terabyte)
1,024 TB	=	1 PB (petabyte)

Why 1,024 and not 1,000? The microprocessor's eight bits forming one byte is actually a binary mathematical foundation. The single bit 0 represents one state or when the bit is off. The single bit 1 represents the other state or when the bit is on. A single bit then can represent two conditions, either on or off. Counting in the binary system starts with 0 and goes to decimal equivalent of the 32 possible binary states in an 8-bit byte that are then represented as hexadecimal values:

00, 01, 02, 03, 04, 05, 06, 07, 08, 09, 0A, 0B, 0C, 0D, 0E, 0F
10, 11, 12, 13, 14, 15, 16, 17, 18, 19, 1A, 1B, 1C, 1D, 1E, 1F
- - - - - - -
F0, F1, F2, F3, F4, F5, F6, F7, F8, F9, FA, FB, FC, FD, FE, FF

Personal Computer microprocessors process 64-bit wide data. That's enough binary and hexadecimal mathematics for today.

Cooling Fans

To help dissipate heat generated by the electronic components, fans are used to pull in air from the surrounding environment to cool the inside of the enclosure. See Figure 11. Good airflow is a very important factor in the operation and longevity of the PC, and a potential source for the accumulation of dust and dirt buildup that can occur on the components. Fans are also the main source of noise.

Figure 11: Typical PC Cooling Fan

Other than in a fanless PC where there are no fans (I could not resist pointing that out), there are often several fans in use:

1. The chassis cooling fan that moves air from the back of the enclosure grill over the motherboard and out towards the front of the enclosure. Often there are two fans, one for pulling air in from the rear and another to push the air out through the front.

2. The CPU cooling fan, mounted on a CPU heat sink, which is

mounted on top of the CPU chip, cools the processor and prevents it from overheating. The CPU cooling fan usually vents to the side of the PC.

3. The power supply fan, which moves air in from the rear power supply grill and into the enclosure.

Suggestion 1: *Vacuum the dust from the outside of your PC*

As air moves into and out of your PC, dust will accumulate on the outside chassis, around the air grill holes. Both the power supply air intake grill, located on the back of the PC, and the CPU exhaust grill, usually located on the left side panel of the PC, will indicate to you how much dust might also be building up on the inside of the PC. I suggest periodically using a vacuum cleaner to get that dust off the outside of your PC.

In order to clean the dust out of the inside of your PC, you will need to disconnect all the cables and connectors, and take the computer outside to blow the dust out. Before you jump ahead and do this, please read the section on "Shutting Down Your PC" and then Removing Dust on page 55.

Storage Controllers: Are chips that control the transfer of data between the microprocessor and the storage device. Hard drives, CD/DVD drives, and floppy disk drives (kind of obsolete) are all storage devices. The hard drive contains all the essential software that turns your computer hardware into a functioning PC when you turn the power on and boot it up. In computer geek talk, a PC when it is turned off and can't be turned back on is called a brick, as it serves no useful purpose in that condition.

I/O Controllers: Are devices that interface between the computer and the human (you). They are called I/O for input and output, or they are sometimes referred to as human interface devices (HID.) I/O devices are things like the keyboard, mouse, trackball, visual display monitor, audio speakers, and microphone. In a modern PC, most I/O devices use the universal serial bus (USB) standard. In a mini-tower

PC you will often see six or eight USB ports for connecting peripheral devices. When you plug a USB device into a running PC, the operating system will in most cases automatically recognize the device and then load the correct software driver for that device.

The letter U in USB means universal, which implies you can plug any USB device into any USB port. The operating system will detect what type of device was plugged in and load the appropriate device driver for it to work in your PC.

THE HARD DRIVE

The hard drive is the bridge between your computer's hardware and the computer's software. It is a device that stores the operating system, application programs and data, and the location where photos, video, music, and other data is stored. You will learn how to see this information in a graphical display in the Windows Explorer section on page 75.

Hard drives are capable of writing and reading data from a spinning magnetic platter inside the unit (or in the case of solid-state hard drives, from static RAM inside the device). Data is transferred from the hard drive into your computer's memory where it is used. It might seem like your PC is executing a program, but that term in computer-speak means to run the program. Programs that use data will read (restore) and write(save) that data to the hard disk for later access.

The space occupied by programs and data (which includes photos, videos, word documents, and more) will eventually fill up the capacity of the hard drive. Please refer to the section on page 70 where we discuss how to properly uninstall programs and delete data from your hard drive.

Where Is the Hard Disk Drive Located?

The physical hard disk drive is mounted on a shelf inside the chassis designed to hold a couple of disk drives. A cable from the power supply, and a data cable from the motherboard, connects to each hard drive. Please refer to Figure 12: Disk Drive Location.

I use the term physical hard drive because the disk controller recognizes each hard disk device as a physical device, starting with device 0. You can add more hard drives, and each becomes the next physical device in sequence. The driver letter designators such as C:, D:, etc., are assigned by the operating system. The drive letter is the virtual drive identifier symbol as a physical drive unit may be divided into multiple virtual drive letters. So virtual drives C:, D:, and E: could very well be on the same physical hard drive.

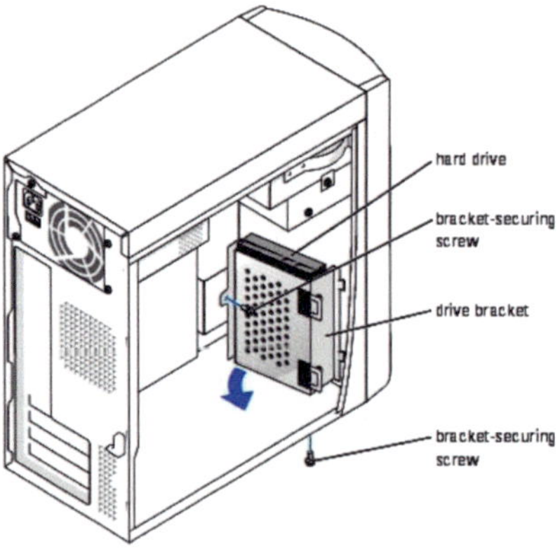

Figure 12: Disk Drive Location Figure 13: Typical 3.5" Hard Disk Drive

Figure 14: Various Floppy Disk Sizes of the past

What Does Computer Speed, Memory Size, and Disk Size Mean?

The computer specifications and marketing literature talk about the various sizes that describe the PC they want to sell you. I often get calls from people who want to know what's the best processor speed and amount of memory to purchase. I usually ask them, "What do you want to use the computer for?" For browsing websites and checking host-based email (Gmail, Yahoo, Hotmail, etc., are considered host-based), a slower, smaller computer is really all you need. Microprocessors come in different speeds: the number of cores or CPU's builds onto a single chip; on-board cache and display memory and other CPU functions.

Memory comes in different storage capacities and different speeds of transferring data into and out of its memory. A single memory chip is available that stores 256KB of information. These chips are mounted on a circuit card in a configuration that offers 2GB or larger of 64-bit memory. Keep in mind that when the power is turned off, the contents of the SRAM (static RAM) are gone.

In the old days of the original IBM PC, two kilobytes (KB) RAM was considered the cat's meow. Today computers come standard with four to eight gigabytes (GB) of RAM.

Again in the old days of the original IBM PC, the movable or rotating storage device was a 5.25-inch floppy disk. It had a storage capacity of 360 KB. Today's computers come standard with 800GB to 1TB or more of hard disk. As you can see, today's computers are *not* your father's (or your) PC any longer.

USB PORTS

The USB ports on a computer or laptop are used to connect USB devices to exchange data between the computer and that device. There are often as many as six or more USB sockets on a PC and two or three on a laptop. Each USB socket provides +5V power and ground to the device, along with the signals for exchanging serial data. There are now four USB standards that have evolved: USB 1.0, 2.0, 3.0, and 3.1—in addition to the new USB-C connector. Each of the USB port types are rated for different data transfer speeds and current delivery capabilities.

When a USB device is plugged into the PC, the first part of the connection process is for the PC to identify what that device is, and load the appropriate device driver for the exchange of data to take place.

Mostly, people plug their mobile devices into the USB port to download pictures from their phone to the computer. They also use the computer's USB port for the purpose of charging their USB mobile device. That's all wonderful, with some exceptions. If you are using your laptop to charge a mobile device, remember you are drawing down power from the laptop battery. It is generally better to recharge devices using an external charging adapter. That includes not recharging small portable emergency batteries from your USB port. I've seen people plugging their e-cigarette device into the USB port of a laptop to recharge it. Don't.

During the hot summer months I've heard of people using a USB-powered fan to keep cool. If the fan's electronics for some reason are not properly designed, it can cause the USB port in your computer to malfunction or burn out. Don't use the USB port to recharge batteries.

Suggestion 2: *Only plug USB computer peripherals into your computer.*

The Software:

You're probably familiar with the term *software*. Software is used to generically describe anything in your computer that is, well, soft. Software is often used to describe the programs that comprise the operating system of the computer. But more often it refers to computer programs written to perform specific functions or that are application programs. Software is soft in that you can't drop and break it! It is not physical. While software can't necessarily be dented, worn out, or damaged, you can delete it, mess up the configuration settings, and certainly mess up the data or database the software uses to perform its functions and remember your previous settings.

When someone asks about your software, you can now ask them for specifics. Do you know which operating system you use, what desktop programs you have installed, or are they asking about the BIOS? Let's further break down the term software and look at what software can be:

The Operating System (OS)

For personal computers, the three most popular operating systems include Windows, Mac OS and Linux. The operating system is what transforms your box of electronic components into a personal computer. It gives this otherwise inanimate object its life; it imbues it with a personality and is the recipient of all your time, attention, dedication, loyalty and source of frustration. You often become a slave to your operating system's quirks and settings. Trying to find where the operation system manufacturer moved the options you knew were there in the previous version of the OS is a very real source of frustration.

Once you get used to working with one operating system version, you might have a hard time navigating your way around a different one. When Microsoft moved from Windows 7 to Windows 10, many people found it difficult to make the switch. Believe it or not, all the operating systems do pretty much the same things; they just have different user interfaces and methods for you to navigate to get those tasks accomplished. Be patient. You will come to love your operating system and all the improvements they made.

The operating system is usually written and optimized to run with specific hardware architectures. While both Windows and Mac OS systems are designed to work with the same or similar Intel family of microprocessors, they are different in their architectural hardware design. Therefore, a program that will run on a Mac will not run on a Windows PC.

As a note, mobile devices like the Apple iPad and iPhones use the iPhone operating system or IOS, and Android tablets and mobile phones use the Android operating system.

The operating system resides on the primary hard drive (or solid-state drive) and is what gets boot loaded to run when your computer power is turned on.

Application Software

Any software that provides some kind of productivity is considered an application software program. This may include Microsoft Office, which is a suite of software programs used for word processing (Word), creating spreadsheets (Excel), a contact and email manager (Outlook), and a presentation program (PowerPoint). Application software for graphics design may include Adobe Photoshop, Adobe Illustrator, and Adobe Premier for editing movies (over the years Adobe bought up most of the competition, creating a monopoly). Microsoft's Movie Maker was once popular, and free.

Movie Maker was officially removed for download on January 10, 2017. Like Windows Photo Gallery from Windows Essentials, Movie Maker is now replaced by the Microsoft Photos App included in Windows 10, which includes Video Editor (formerly Windows Story Remix).

Accounting software like QuickBooks desktop, for example, is an application program, and so is any other line of business program that is used to run and manage your business.

What may not necessarily be considered an application software program is anything that does not reside directly on your computer's hard disk, such as cloud-based websites. The web browser is the window to the website you are looking at. Chrome, FireFox, and Edge are examples of web browsers. You interact with the website using a web browser. These are called web-based programs, or more often just cloud computing. The website contains the application that you access from the website, and the web browser runs the JavaScript language client-side programs that the website downloads to your web browser, to imbue it with nice graphics and user functionality.

When Things Work Correctly

British science fiction writer Arthur C. Clarke formulated three adages that are known as Clarke's three laws; my favorite is: "Any sufficiently advanced technology is indistinguishable from magic." I've adopted a version of this adage in my consulting practice to read "When technology is properly implemented, it is barely discernable from magic." We know from experience that when our PC is not working well it can create a terrible nightmare for us. Most often when our computers are not working correctly, it is because we are not working correctly with them.

To help us prevent our technology from not working the way it should, and to avoid the nightmares, we'll start to look at how the hardware and software work together to make a functional working personal computer. By understanding how each of the hardware and software components function in harmony, we can begin to take steps to keep each part of the computer in tune. Much like an orchestra, each instrument plays an integral part in making a great symphony. Each of the parts of your computer must play in tune with one another. So let's start up the orchestra.

TURNING ON THE POWER

When your PC is turned off, a part of the power supply is actually still on (assuming it's plugged into the AC outlet), providing a small amount of voltage to the power push-button circuitry. When you press the power button, it electronically signals the power supply to turn on the other voltages that in turn starts up the PC. Many power supplies have an additional hardware switch in the back of the power supply near where the power cord connects to the power supply. This switch will turn off all power to the entire computer and is usually left on all the time.

When you press the power button on the PC, the first thing that happens is the various power supply voltages turn on. These voltages then feed into the motherboard where the microprocessor begins to run and the peripheral devices start to power up. In particular, the ROM BIOS runs its boot program to make sure the basic hardware is ready and that the hard drive motor spins up. Once the ROM BIOS has determined that it is okay, it will then start to load the operating system.

There are two terms that are used to describe how to boot your computer: the first is a cold boot, which is when the power supply is off. The other is a warm boot which is essentially resetting the CPU logic but not turning off the power. In a cold boot, the memory (RAM) and all the peripheral devices (hard disks, USB devices, keyboard, and so on) are started when the power comes on. During a warm boot, power to these other devices is unchanged. I'll explain how knowing that difference can be useful in diagnosing a problem.

Suggestion 3: *NEVER turn off the PC without first going thru the operating system's shut-down process.*

THE ROM BIOS

From the moment you push the power button until the sign-in screen appears on the display, there is a whole lot of stuff going on inside your PC. The first start up process is when the microprocessor runs or executes the ROM BIOS (or read-only-memory basic initial operating system). The BIOS is specially written software designed to support the hardware architecture of the various types of components designed into the motherboard.

The ROM BIOS is usually written by the manufacturer of the motherboard to match the components they used in making this motherboard. It is a complex piece of software that boots up the computer hardware by sequentially testing to make sure each electronic component is working correctly. It checks to make sure the memory is functioning, and it checks to see if a keyboard and mouse are connected. You might notice the Num Lock and Caps Lock LEDs flashing as they are reset, and the DVD reader door will close if it is open. The BIOS checks for the presence of a display monitor, then clears the display memory and shows the initial booting information. It then checks the chassis temperature, turns on the fans, and makes sure they are working at the correct speed. It checks the temperature of the CPU, and then it performs a very long list of other critical steps.

It's like mission control at Kennedy Space Center getting ready to launch a rocket into space—once all systems are go, your computer operating system will then load from the hard disk. If any single part of the system is not a go, your computer will not start. Often you might see a cryptic message displayed on the monitor telling you, or someone who knows, what went wrong with the initial countdown.

I was the software development manager at Eagle Computer back in 1982, working with the BIOS development team. I can assure you it takes a team of really smart people and long hours to write the code ensuring each and every hardware component is tested and working correctly during this startup boot phase. What is amazing to me is that after the thousands of hours that go into designing and debugging the BIOS, it runs from start to finish in just seconds. That is a humbling experience for a programmer.

LOADING THE OPERATING SYSTEM

Once the BIOS program passes all the essential initializations and testing, the microprocessor is then cleared to load the OS program from the hard disk. The OS is copied from where it is stored on the hard disk and loaded into the SRAM, where it will then bring life into the hardware. In a Windows PC, you will see a screen indicating that windows is starting. While you watch the flag wave, you will now appreciate all the stuff going on inside your computer while you wait.

The OS is what makes your computer a personal computer. The OS is comprised of a bunch of software programs or modules that provide specific functionality that form the operating system. There are many, many hundreds of software modules that make up the OS. The loading of these modules is hidden from view. If you want to view a list of what drivers and components are running in your PC, you can use the Task Manager that is part of the OS. Right-click on the Windows toolbar, select Start Task Manager then select Processor or Services tabs.

Once the operating system is loaded, key services are started and your computer is now ready for your first command. These services might include connection to the internet via a connected ethernet, or via Wi-Fi to your network. The settings are loaded to initialize the internet connections which you set and saved the last time you configured your network settings. These settings, and every other setting you customized before, are magically brought back to life and loaded into memory from the hard disk where they were stored. Other settings might include loading your printer setup, retrieving a photo you have as a background image to be displayed on your monitor, and loading the various drivers for all the peripheral devices you might have connected to the USB ports. There is a lot going on when the operating system boots up and is fully engaged.

SIGNING IN TO YOUR COMPUTER

Now that your operating system has loaded, it is waiting for your wish to become its command. The first wish is that you sign in using a username and password, a PIN, or a fingerprint authentication. If you don't have a username and password sign-on page as a guardian to your personal computer, you should. It is there for your protection. Everything you would like to do with your computer from now on is going to involve some kind of software. Start up a web browser to check the weather: it loads the browser application into memory. You launch your email program and download email from your server to the PC: it loads the email client application into memory. If you want to listen to music while you are working, it loads an MP3 media player into memory, then it streams the digitized music from your hard disk or online provider into memory to be converted to sound and played by the audio chips located on the motherboard and delivered to your speakers, headphones, or Bluetooth connected listening device. It's amazing how many steps it takes for the computer to perform what appears to be a simple task like listening to music. It has taken decades for this technology to evolve, and it keeps getting better with each new and updated revision of the technology. To make something simple usually takes a long and difficult process of revisions, changes, and evolutions. Why can't they just do it right the first time?

Any application program that you have installed on your hard disk is now waiting for you to launch, which then loads it into memory to start running.

Often people say their computer is running more slowly because they have too many programs or too many photos on their computer. No. Not really. Programs and data are only stored on the hard drives for future access. Only when they are loaded into memory are they available to be accessed by the microprocessor. Having a lot of stuff on your hard drive does not make your computer run more slowly. The software programs on the hard drive do not weigh down the disk either—they are electrons.

In the old days, the size of RAM was at a premium cost, but the application software took up more memory than was available to run. The operating system would create a swap file on the hard drive where portions of the software program could be moved into and out of memory. Then, yes, the computer program would run more slowly because the CPU was busy swapping it into and out of memory. More memory means less or no swapping and therefore a faster computer.

WHAT IS THE MICROPROCESSOR DOING?

In a healthy computer, that is, when the operating system is running as it should and there are no other programs hijacking your computer (aka Trojans and viruses), your processor should be idle for the most part. The input/output drivers are running their code, listening for the keyboard to send keystrokes; the mouse driver is listening for changes to the optical sensors on the table, instructing the cursor on the screen to change position; and the display drivers are updating the visual representation of the cursor location. After signing into your computer, and after all the operating system programs are loaded, your computer should be doing not much of anything else. If it is, then this might be a sign that you have virus or Trojan.

TASK MANAGER—LOOKING INSIDE AT THE OPERATING SYSTEM

It's time to take a deeper look into what is going on when your computer is sitting seemingly idle, waiting for your commands. The Task Manager is a window into what is going on with the processes, and it comes with the Windows system as a tool for you to use. Learning how to open this tool to peek inside the microprocessor functions is a big, but simple, first step in your understanding of what your computer is all about. This will help you remove any mysteries as to what your computer is doing, and it could assist you in preventing and diagnosing problems.

The Task Manager is my go-to tool when I think my computer might be running slowly. I open up the Task Manager to see what's running and try to determine what might be taking up the CPU resources that could be slowing things down. Often, the operating system is just performing routine cleanup processes and housekeeping tasks in the background when the computer is not in use. At other times another program may have started that I didn't know about, and it is sending information out to the internet, sometimes to unknown destinations—a sign of a possible virus. It could also just be Skype automatically connecting to their servers, or an email program checking for and downloading email. I like to know what is running in my computer. Over time you will begin to recognize the programs that should be running all the time, and you will eventually learn to spot programs running that should not be running.

As with most Windows tools and programs, there are number of ways you can launch a tool. From the task bar along the bottom of the screen, you can right-click, then select Start Task Manager. Alterna-tively, click on the Windows icon on the bottom left side of your screen or press the Window button on your keyboard. Then from the menu bar search window you can search for Task Manager. If you are really desperate to find it, try the Ctrl-Alt-Del key which is often used when things are hung up. From there you can start the Task Manager as well.

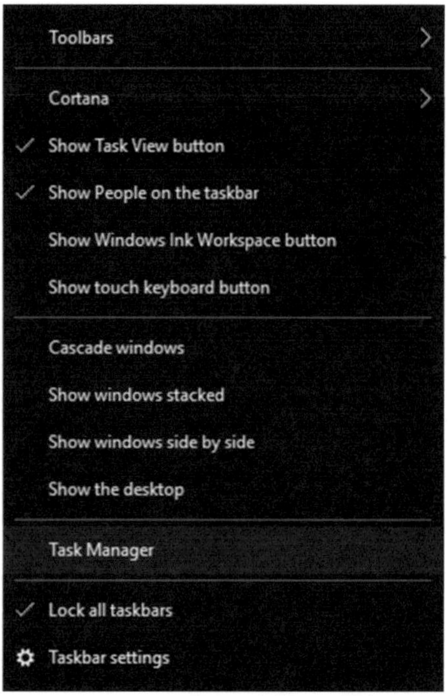

Figure 15: Right-Click to Start Task Manager

We'll take a closer look into the Task Manager later on page 58.

What Can Go Wrong? How to Prevent It

Now that you have a reasonably thorough understanding of the many hardware and software components that comprise your personal computer, you can begin to look at those things that can have an adverse effect on each of those components. The ultimate objective to having this understanding is so that you will use your computer with more respect and take better care of this important thing in your life. With this perspective you can begin to use your computer with an understanding of what to do, or not do, that will lead to a longer, more productive, and healthy life span for your PC as well as for yourself.

EXCESSIVE HEAT

Computers require 120 volts of alternating current electricity (VAC), and through a switching power supply it is converted into the various direct current voltages (D.C.) and regulated current needed by the various components. The conversion of power, while fairly efficient in today's well-designed power supply, still loses some efficiency as a result of the conversion into heat. Since the power supply is one source of heat, the power supply enclosure has its own cooling fan built into the enclosure.

Integrated circuits are also not one hundred percent efficient devices. Making a near one hundred percent efficient semiconductor has been the holy grail of superconductor research. The semiconductor materials used to make the chips contain a small amount of internal resistant; this resists the flow of electrons, which is then converted to heat. The CPU chip loses some of its efficiency and is converted to heat as well, and since the CPU is the biggest chip on the motherboard, it is the single most heat-producing component on the motherboard. As the CPU is also the critical component doing all the processing, it's important to keep the CPU running cool. The CPU, in all likelihood, has its own cooling fan mounted on top, keeping the air flowing over the surface, removing excessive heat. Keeping the air flow moving through your computer is important.

Airflow is affected by a number of things, including the place in your home or office where your PC is located. Placing it on the desktop or on the floor is great, but make sure you don't block the air intake fans on the back or side as airflow might be restricted. Never place the chassis inside a desk drawer, a cabinet, or other small space, unless it has suitable ventilation that permits good air flow.

A point I want to make is if you do place your PC on the floor, try to elevate it at least 2" to 3" above the floor. Why? Dust tends to settle on the floor, which can be sucked in through the cooling fans, increasing the amount of dust your computer collects. At my home office, my mini-tower PCs sit on a homemade stand I made in the workshop. The PC stand is made from a piece of plywood measuring about 8" by 18" (depending on the size of chassis), and two 2'x4"x18" pieces of wood that serve as the base.

The ideal temperature for your computer environment is between 60°F and 75°F. The cooler the room, the better. The inside operating temperature of the electronic components can reach 96°F, so having cool air to offset the higher inside temperatures is essential. As the room air temperature increases above 75°F, the ability of the PC to keep cool is reduced. So, why is temperature important?

TEMPERATURE INDUCED STRESS ON COMPONENTS

Referring back to the motherboard, it is a printed circuit board that is made up of between four to eight layers of very thin, insulating fiberglass material and very fine, conductive copper patterns that electrically connect the pins of the chips forming a functional system. The integrated circuits, or chips, are typically made of plastics which contain the semiconductor electronics referred to as the packaging. Chip manufacturers rate the operating temperature range of their parts based on the type, quality, and application or environment the chips will be suitable to be used in. The chips that are sent into outer space, as you might guess, have a somewhat different operating temperate range than the chips used in your PC. They cost a whole lot more too.

When semiconductors and the package containing them operate, they heat up to their normal operating temperature. When the power is turned off, they cool down to ambient room temperature. In the process of heating up the material expands ever so slightly. As they cool back down they will contract ever so slightly. This is called thermodynamics. Any object when it undergoes temperature change will expand and contract as it heats up and cools down. This is the single biggest enemy of your PC—thermal stress.

Turning on your computer from standard room temperature of 20°C (68°F; temperature in the electronics world is measured in Celsius) will quickly warm up to a power-on operating temperate of about 30°C (86°F). The exact operating temperature of your CPU and other electronics will be different between manufacturers. The change in temperature from room temperature to operating temperature causes the chip material to expand from a cooled condition, and it causes a very small amount of stress on the component. Equally, when the PC power is turned off, the components cool back down to room temperature, again inducing a bit of thermal stress. With enough cycling of the power and therefore cycling of the temperature, any one of the components on the motherboard might eventually fail. I say *might*, because silicon die and packaging materials have improved significantly in the past twenty years, but thermal stress is still a factor in component failure.

Suggestion 4: *Leave your computer on. Do not shut it off.*

If you use your computer on a daily basis, my recommendation is not to shut down the computer. Leave it powered up. This will keep the temperature of the electronics at a fairly constant operating temperature range that will minimize the stress that can lead to chip failure.

Rather than shutting off your computer, you can simply log off or sign out from the operating system software. You might also leave your LCD monitors turned on. Your operating system has a monitor timeout setting and newer LCD displays also have an automatic power-save mode built into the monitor. This applies to laptop computers as well.

If you think your PC is not getting sufficient air flow and cooling, there are some free software tools that will help you peek inside the operating hardware. A program called HWMonitor from CPUID.com will give you a comprehensive live display of your hardware health. With this or similar monitoring programs, you can see the clock speed of the CPU cores, temperature of each of the major components, and even the speed of the cooling fans in RPM.

DUST

Closely related to temperature and excessive heat buildup inside the PC chassis is the accumulation of dust. Dust particles are pulled in by the various cooling fans, and over time dust settles on the components on the motherboard. Dust on the components contributes to heat buildup, as cooler air flow cannot reach the surface of the semiconductors to remove the heat. Dust can also buildup inside the fan motor, causing the motor bearings to wear out. If your computer is squealing, it is most likely coming from a fan that is slowing down. In fact, if your computer is making any noise, it is definitely coming from a fan, as it is the only moving part in the PC (not including the motor that rotates the hard disk.) Most PC replacement fans are available in standard sizes and are cheap and easy to replace.

Spring cleaning is a great idea for every computer in your home or office. Depending on the environment, you might want to blow the dust out more often. I have a client with an office located inside a mobile trailer, situated on a farm surrounded by fields of dirt. Other than a minicomputer located on an airport runway controlling the runway lights at an international airport, I've never seen more dust build up inside a PC than at this farm location. Hopefully your home is not as dirty and dusty as that, but even in my otherwise clean home office, dust can be seen building up on the outside of fan intake grids.

There are several methods to clean the dust out of your PC and keyboard. One of the easiest things you can do is to keep handy several cans of compressed air, used to dust electronics. I use the pressure from the compressed gas to clean out the dirt particles that somehow manage to get into my keyboard. I keep a can on my desk, and once a week or so I blow the dust out of the mouse and track ball as well.

Cleaning the dust out of your computer is a bit more involved, so the procedure is described in detail in the Removing Dust section starting on page 55.

Suggestion 5: *Clean the dust from the inside of your PC*

Figure 16: Compressed Air Can

You might notice the suggestion about cleaning dust repeated a few times. That is because performing a spring cleaning of your PC can really make a difference in how long your computer will continue to operate without developing hardware issues.

POWER TO THE COMPUTER

Providing a clean source of power to your PC is just as important as any of the other aspects for your computer's good health. Power is often an overlooked factor, because it's so easy to just plug your computer into any wall outlet, turn it on, and get it to power up.

What do I mean by clean power? It basically means power that does not change or vary from the normal range of voltage and current. If you share the power outlet with a refrigerator or freezer, as an example, the quality of power to your computer will vary when the refrigerator compressor turns on and off.

Suggestion 6: *Never plug a computer into an outlet that shares a power circuit with a major appliance.*

Although the power supply inside the PC does an excellent job at regulating and maintaining consistent voltage to the components, things like power surges can get through the circuitry and cause parts, including the power supply itself, to blow out. Power surges are much more common than you might imagine, as they are not always noticeable. As other electrical things inside the home or office are turned on and off, they can cause rapid, short changes to the alternating current, affecting the whole house.

A blackout is when total power is turned off to an area. It can be intentional or unintentional and can last from a few minutes to a few hours. A blackout can be caused by power line equipment failure caused by a storm or an accident, or by the electric company shutting down the power to do work on the power lines. The power company usually provides advanced notice for planned outages.

A brownout on the other hand is an intermittent reduction in the voltage level. The voltage level of the A/C line could drop below the normal and safe voltage operating range of your computer and can have serious damaging consequences to the electronic components in your computer. A brownout often lasts just a few seconds, or can persist from a few minutes to several hours. Brownouts can cause damage to any and all office equipment and home appliances, because the reduction in power puts motors and electronics at very unstable conditions. A brownout can harm TV sets, microwave ovens, and even refrigerators, if they are on when the brownout occurs. If you don't take proactive steps to turn off sensitive electronics during a brownout you can be in serious trouble, quickly.

If there are thunderstorms in my area, which don't occur as often as I would like in Southern California (I love a good thunder and lightning storm), the first thing I do is make sure the TV and any radios are turned off. I then unplug any electronic device that is connected to the A/C line. A good lightning strike to a power transformer can send dangerously high voltage and surges into your neighborhood through the power lines and into your house. While it isn't likely to happen in Southern California, I don't like playing those odds with my new 65" HD TV and certainly don't want to risk losing the information on my computer.

There are a few things you can do to protect your electronic devices and personal computer against the inevitable power surge, brownout, or blackout:

Surge Suppressors:

As the name implies, a surge suppressor suppresses surges occurring in the A/C line power and keeps them from getting up into your computer's power supply. It filters out sudden increases or spikes in voltage that might occur. I often use surge suppressors in the bathroom where I keep my electric toothbrush and shaver plugged in for recharging. I use them on the power outlets where I plug in my mobile phone USB chargers.

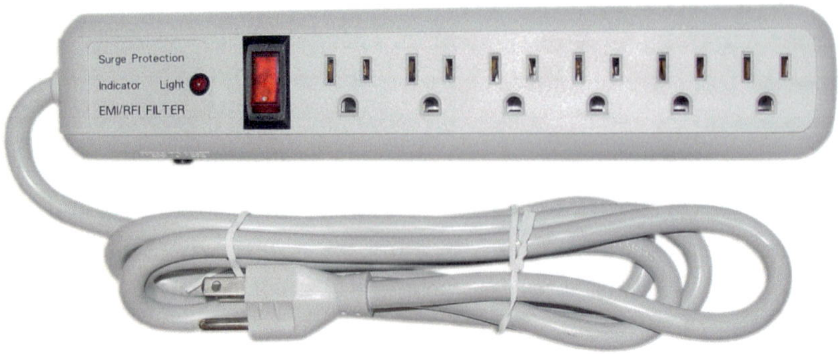

Figure 17: Surge Suppressor

Battery Backup:

When was the last time you experienced a power outage? In some parts of the country power outages are a more common occurrence than others parts of the country. But one thing we all know is the when the power goes out, everything that runs on electricity stops working. When I say stop, I mean it shuts down in the middle of whatever you were doing. If you are doing work on your computer, this could mean losing a day's worth of your time. It could mean time and effort and your thoughts that went into writing that letter are now gone. Having the power go out suddenly also means the hard drive in your computer might not have turned off gently, possibly causing disk errors or hard drive failures.

In my home office I have several 1,200KVa and 1,500KVa [see Figure 19] battery backup units; one is used to power my cable modem, routers and voice-over-IP equipment, or my communications devices. Another powers my main development computer and monitors. Another smaller unit powers my digital PBX computer and an HDTV monitor so that I can tune in to watch the news and find out what is going on with the outage. I have a battery backup device in the living room to provide power to low-current LED lights. When the entire neighborhood is blacked out, my house is all aglow.

Figure 18: Battery Backup (350VA UPS)

Figure 19: Battery Backup (1500KVa UPS)

Suggestion 7: *Plug your PC into a UPS battery-backup device*

Selecting an uninterrupted power supply (UPS) battery

There are several name-brand battery backup devices in the USA, including APC, CyberPower, and Tripp Lite. Even Amazon sells an Amazon Basics battery backup device. Most of these devices use a standard 12V industrial sealed lead-acid battery. The batteries can last two to three years before they need to be replaced. Checking the remaining life of your UPS battery should become part of your good habits. It's easy to open the UPS and replace the battery but be careful—these batteries are heavier than they look.

When making a selection, check to make sure the UPS device has a serial USB connection to the PC and comes with UPS power management software. This is important, because if the UPS battery should become drained before the mains power has been restored, the UPS will just stop delivering 120VAC and your computer will shut off. The USB connection and power management software to your PC will signal your computer that it's time to shut down gracefully, initiating the shutdown procedure just like when you shut it down using the menu function. After you install the UPS and power management software that comes with your UPS device, test it. Make sure it's connected and that the UPS settings will shut down your PC. Some business edition power management software can be configured to send you a text message when the power in the house or office fails.

How much battery backup power do I need? The higher the current or capacity that the battery backup unit can deliver, the more devices you can power and the longer the UPS will continue to generate the needed 120VAC voltage output. Some of the better UPS devices have a display indicating the power loading along with your AC input voltage and frequency (it should be around 110V at 60Hz). As a general rule, you can calculate what size UPS you need based on the formula:

1.6 * Wattage Load = Minimum Volt-Amperes (VA)

A typical desktop PC consumes about 400Watts, so you could get by with a 640 VA UPS device. Before you go out and buy a 640VA UPS, continue reading.

How long will the UPS provide power to my PC? That calculation is more complicated. When making a selection you can refer to the manufacturer's specifications and guidelines. Since I connect only a few devices to one UPS (a single PC and a monitor or two) I like to have the extra power to keep my PC running longer. During a power outage my 1,100VA UPS will keep me going for about 40 minutes. Your actual performance may be different.

Suggestion 8: *Check your UPS Battery periodically and replace it when it is getting weak.*

THE DISK DRIVE

The operating system and application programs and data, which includes images and audio/video files, all reside on the hard drive of your computer. The amount of storage space available on your PC is based on the size of the hard drive installed in the computer. You can quickly see the total hard disk space you have available and how much is currently in use. To see that information, open File Explorer, select the primary disk drive (usually C: drive), then right-click and select Properties.

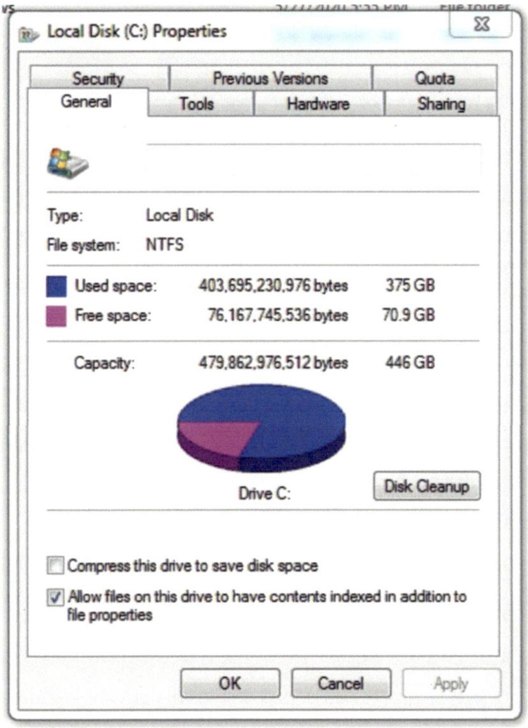

Figure 20: View Remaining Disk Space

HARD DISK ERRORS

Probably the single most critical part of your computer is the hard drive. This device holds the operating system that is booted into RAM when the ROM BIOS completes its hardware check. This starts your computer. The hard drive contains your software application programs. It stores all your documents, spreadsheets, photographs, movies, music, and just about everything else you use the computer for. The safekeeping of this data is paramount to your happiness and peace of mind.

A standard 3.5" hard disk spins at 7,200 revolutions per minute (RPM) . Calculating the circumference of the disk platter in inches times the RPM, the distance the outside of the hard disk travels in one minute is 79,200 inches, which is also 1.25 miles per minute, 75 miles per hour, 1,800 miles per day, and 657,000 miles per year. The only moving parts in your computer are the rotating hard drive and the fans. No oil change required.

HOW CAN HARD DISK ERRORS OCCUR?

There are a number of ways hard disk errors occur. As you understand how each part of your computer works together with the other parts, you will learn how to reduce the risk of hard disk errors.

One way a hard disk error can occur is if you suddenly lose power to the computer. Using a battery-backup device can help protect your computer and the data on your hard drive. The other way to cause a hard disk error is by shutting off the power supply using the power switch on the back of the power supply, or as I've often seen, switching off the power strip. Always use the shutdown procedure if you are going to turn off the A/C to the computer.

THE DREADED BSOD

I wanted to mention a situation that fortunately does not occur that often. As computer hardware has become more reliable (e.g., better quality RAM and hard drives), this condition is seen less frequently.

Blue screen of death (BSoD) refers to the notorious error message displayed in Microsoft Windows operating systems. Because the BSoD message indicates an unrecoverable system crash or freeze and leaves the user no option but to reboot the computer, the blue screen has been both feared and ridiculed by Windows OS users ever since its introduction through Windows 3.1 in March 1992.

I'm hoping the material in this book will help you avoid the BSoD. Recovering from BSoD might be as simple as rebooting your computer, as the error message might suggest. But the BSoD might also be an indication of something more serious. If you ever get the blue screen don't panic. OK. Panic. Then, get over it. First, start calmly and from the beginning: turn off the power from the main power supply switch. Wait a moment and then turn it back on. Pause another moment, then press the power button on the front panel. Watch the monitor carefully, and see if you can read any additional messages that might help to pinpoint the source of the problem [see Figure 21.] In some cases, the error will correct itself when all the components are reset—often times it won't. If it doesn't recover, then you can panic.

A few years ago, I inserted a USB thumb drive into the back of my PC in order to transfer files. I left the thumb drive plugged in and forgot it was there. A few weeks later I shut down my PC, and when I restarted, I got a BSoD telling me there was no bootable operating system. After a brief panic, I rebooted again, this time noticing that during the boot process the BIOS monitor indicated it was booting from a USB drive. Remembering it was in the back USB socket, I unplugged the USB thumb drive and voila, my computer started up normally. Panic over.

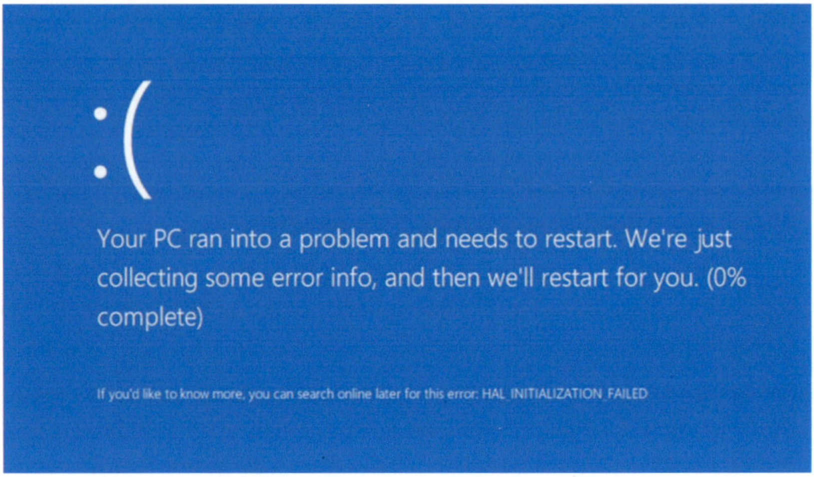

Figure 21: Blue Screen of Death

BACKUP YOUR DATA

You are learning to take steps to insure a long and healthy life for your personal computer. However, there are still circumstances that arise that are beyond anyone's control, despite all the precautions you may have in place. Some of these circumstances include:

- Your hard drive just decides it's time to fail—while they are built to last they do contain moving parts, and electronics, that can fail. A hard disk crash usually means your data is not going to be accessible.
- A power failure—and you don't have a battery-backup system with automatic shutdown, so the hard drive power going away suddenly might cause errors.
- Unintentionally deleting files or folders that contain your information.
- Inadvertently installing a virus or worse, a Ransomware cryptography virus, which will encrypt all your files.

There are many reasons why your valuable information might become inaccessible or deleted, and having a backup of your data can be invaluable. In fact, doing routine or automatic backups is just as important as any other PC habit. Backing up your information should be on the top of your to-do list every time you complete your work.

There are many ways in which you can back up information. In the simplest, you can transfer important files to a thumb drive or memory stick, or externally-connected USB backup drive. I often transfer Microsoft Word and Excel files or other documents to a thumb drive, so I can transfer them onto my laptop when I need to work while traveling; I also use this method to transfer a video file I created to show at a meeting. This method is okay if you know what files you need to move to another location, but it's not really a backup per se.

The most common method used to backup up information is to connect an external USB backup drive to the PC. These drives usually come with backup software that does a reasonably good job of backing up data. You should make sure to set the backup software to do a full backup with automatic incremental backup on a regular schedule.

Just as important as backing up your data is the ability to restore your data when you need to. Many of these backup programs do what is called file-based backup. In other words, it makes a copy of the original file on the backup drive. Some disadvantages to using an external file-based USB backup drive include:

- Files on a USB drive can be encrypted by Ransomware.
- If your computer is stolen, the thief might also take your backup drive, which is usually sitting next to your computer.

Suggestion 9: *If you use an external USB drive for backup, unplug it when you are not performing a backup.*

The better backup programs, which are also typically not free to install, will compress your data into a format that most likely would not be touched by an encryption virus.

Off-Site or cloud backup is probably your safest bet, assuring that scheduled backups occur, and that you can restore any file, whole folders, and even the entire operating system, should you need it. With most of these cloud-based backup solutions, there is a monthly or yearly fee. A full backup might take a full day to perform, depending on your internet speed. A daily incremental backup, scheduled to run at night when you are not using your computer, only backs up the information that has changed during the day. With a full backup available in the cloud, you can restore a new hard drive, with everything you had previously, in the event the hard drive is not repairable. Trust me, cloud backup is money well spent.

Looking Inside Your PC

For some of you, you might have been anticipating this chapter. For some of you this is what you have feared. This is the part of this story where many of you might get squeamish or start to feel uncomfortable. There is no need for worry or fear. Opening up your computer is not difficult, it is not dangerous, it isn't rocket science or like doing brain surgery. You won't do any harm to your computer. No need for fear. This is why you are here.

"But why do you want me to open my computer?" you may ask. Well, the first reason is so that you can effectively clean out the accumulation of dust that is building up inside your computer. The longer you've had the computer, the more dust you will see coming out. If this is the first time you are cleaning out the dust, you might be in for a big surprise.

The next reason is just because your computer is your friend, and you should become familiar with what is inside the box. My principle is that the more you know about things, the better you will be at understanding what to do in the event something goes wrong with one of the components.

Before opening up the computer's panels, you need to disconnect the peripheral cables from the back (and sometimes the front) of the PC. Before you start to unplug things, first make sure the computer is powered off and then unplugged from the AC power.

SHUTTING DOWN YOUR PC

To properly shut down the computer, select the Start menu on the lower left of the screen. Depending on the version of Windows you are running, you might see a power icon as shown in Figure 22. In either case, select Shut down. If the button says something else other than Shut down, click on the arrow to the right of the button, and you will be presented with some shut down options. Select the Shut down option, which will start the shutdown process. That option will also become the default selection for the next time. Make sure Shut down is clicked.

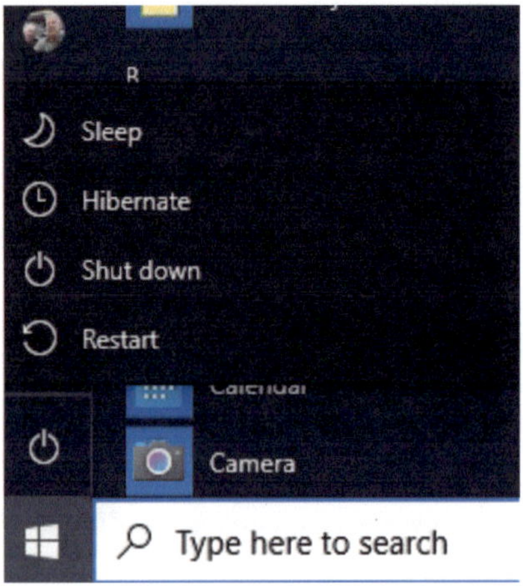

Figure 22: Windows 10 Power Shut down option

Suggestion 10: *Only use the operating system's shutdown process to turn off your computer.*

<u>Never</u> turn off your computer in any of these ways:

- Unplug the power cord or turn off the power strip.
- Turn off the switch in the back of the power supply.
- Hold down the power button for five seconds.

Any of these methods should be used only in an emergency, if something really drastic occurs, like smoke coming out of the computer, or the normal shutdown option is not accessible. Otherwise, you may damage the hard drive or electronics.

During the shutdown process, the operating system will close any programs that you have opened, and it will ask you to save any unsaved work. It will then save your desktop screen information and close out all those drivers and programs you learned about that get started when you turn the computer power on. Once the shutdown is complete, the monitor goes blank, any LEDs on the front panel turn off, and the fans stop turning. This is the graceful way to turn off your computer. Now you can proceed.

DISCONNECTING THE CABLES

The keyboard and mouse cables, along with the cables that connect your printer, scanner, external camera, and other devices, are all USB devices. As the term Universal Serial Bus implies, they are universal, which means you don't have to mark each USB cable, as they do not need to be plugged back into the same USB socket. It doesn't matter where you plug in a USB connector, as long as it is in a USB socket.

When you boot up the computer, each USB device will be associated with its device driver automatically. I've seen people actually number each cable corresponding to a label they put on the back for the PC. I've even seen someone label the power cable. That way when they relocate their computers in the office they get the correct power cable connected back to the correct computer.

The Ethernet connector is your connection to your local area network, also known as the internet. Unplug the Ethernet cable by pressing on the plastic tab and gently releasing the tab and pulling the RJ-45 (internet) connector from the back of the PC. If you are using Wi-Fi, then you most likely will not have the RJ-45 connection hardwired to your local area network.

Loosen the screws on the monitor cable if you are using a VGA monitor, or if you have an HDMI monitor, simply unplug the HDMI cable. If you have more than one monitor, then this is a reason you can get fussy and put a label on the monitor cables. When you reconnect the monitors you can ensure you've connected the left and right monitors.

If you have speakers and a microphone connected to your computer, unplug each of the phone jacks from the I/O panel on the back. Take note of the color code: the green jack is the stereo speaker connection; the red is the microphone jack; and the pink is an auxiliary input jack. The plugs and jacks are usually already color coded.

Finally, unplug the three-prong power plug from the back of the power supply.

OPENING UP THE BOX

Now that all the cables are unplugged and disconnected, it's time to move it to a work space. I usually take mine to the garage and place it on a workbench. After I remove the cover, I'll take the PC outside to blow out the dust. But before I open the PC covers, I usually go back right away with a vacuum cleaner, to where I removed the PC. I like to clean the floor or carpet and the surrounding area of dust. It's amazing what you find hidden behind your PC.

Now, back in the garage or workspace, it's time to remove the cover. Looking from the front of the PC, the main cover is usually on the left side. Some PCs, like Dell, use a latch mechanism that you pull and then slide the cover off, toward the rear. Most, however, have two screws holding the cover on. Using a Philips screwdriver, remove the two screws. Place them someplace out of the way. Small parts have a funny way of disappearing while you're busy with something else. Slide the cover backwards about a half inch and remove it. It probably has small tabs that hold it in place when it is slid back into place.

Now that was not difficult. Take a look inside the enclosure. Referring to the picture below, you will see the various parts and components you learned about in the first chapter the Anatomy of a PC. See any dust?

Figure 23: A Look inside Your PC

In the upper left corner, you see the power supply power socket in the red along with the fan. In the center is a circular air guide with a CPU heatsink and fan that pulls in air through ventilation holes in the cover panel that you removed. Just below the CPU are various components mounted on the motherboard, including the sockets where the RAM boards are installed. On the right-hand side of the photograph, towards the front of the enclosure, is where the storage devices are located. On the top right and facing the front of the PC is usually a DVD/CD player, and below that a floppy disk device, and below that the hard disk is mounted on a sheet metal shelf. In this photograph you see flat ribbon cables that connect the storage devices to the motherboard. These ribbon cables use a technology call IDE. In newer PC's the cable that connects the devices to the motherboard is called SATA. If you ever need to replace or add an additional hard drive, you will need to know what kind of drive your PC has: IDE or SATA.

REMOVING DUST

When you are finished visually inspecting the inside of your PC, it is time to move the computer outside where you will blow the dust out.

If you have a vacuum cleaner where you can reverse the hose into the exhaust port, attach the narrow crevice tool. You can get a nice strong air pressure that is great for blowing out the inside of the computer. I have a 100PSI air compressor in the garage with a blowing nozzle that works great. But before you jump up to blow the dust out of your computer, make sure you are someplace where all that dust can go—preferably outside. You might even want to wear a mask and even some goggles.

I have often invited my clients to come outside with me to witness the dust coming out of their computer. It's often quite a dirty show.

With the air pressure nozzle, move it around, blowing dust from every part of the chassis. I often start by blowing air into the power supply from the inside out. Then blow air from the rear of the power supply back into the enclosure. Move around, as the dust will begin to relocate to other parts of the enclosure. Make sure to blow air onto the CPU fan and CPU heat-sink area. In fact, all of the fans should be a target for dust removal. Blow air from inside the enclosure behind the storage devices and out toward the front of the PC grill. Then go to the front grill and blow air back inside. Repeat this a few times until you've removed as much dust from everywhere possible.

Once you are satisfied that all the dust has been removed, clean the side panel CPU intake holes. Take the PC back onto your workbench and replace the side panel, making sure the tabs fit into their slots. Then slide the cover toward the front. You might need to give a bit of a knock to get it to close completely. Replace the two screws, and voila, a clean machine.

Keeping your PC Software Clean

Just like your PC hardware, the software that runs on your hardware should be kept in tip top shape. In fact, keeping your PC running a long time is just as much about maintaining good software habits as it is maintaining good hardware habits.

WHY IS MY PC RUNNING SLOWLY?

I've often heard people ask "My computer is running more slowly than before. Maybe I have too many files on it? Should I delete some so it will run faster?"

The answer is NO, probably not the reason. The hard drive is a storage device that keeps information organized and available any time you want to access that program or information. The data on your hard drive does not slow down your PC. It's like your bank account and your wallet: your wallet is where you keep a small amount of cash you might need for immediately use. Your bank account can hold as much money as you can put in it and is there waiting for you to withdraw what you need and transfer it to your wallet. You don't keep all your money in your wallet (well, in today's economy, some people might).

There are many reasons why a computer appears to be running more slowly, having to do more work, and not leaving you with enough- machine resources or time to do your work.

SYSTEM PROGRAMS

One of the things that might slow down the computer operating system is having too many installed programs that load when your PC boots up. These programs run in the background, meaning you don't see them or know that they are running and doing things that take up processing power.

Most of the background programs that get loaded automatically are doing tasks that are essential. The device drivers are software programs written by the particular hardware manufacturer that interface between the various hardware devices and the operating system. Every hardware device in your PC uses drivers that are loaded to run, and each is essential to interface with the operating system.

APPLICATIONS, PROCESSES, AND SERVICES

Before we dive deeper and look at the various software things that are running in your computer, let me first explain the three major types of software that run on a PC:

A **service** is a process which runs in the background and does not interact with the desktop. In Windows, services almost always run as an instance of the svchost.exe process, the windows service host process; however, there are sometimes exceptions to this.

A **process** is an instance of a particular executable (.exe program) file running. A given application may have several processes running simultaneously. For example, some modern browsers such as Google Chrome run several processes at once, with each tab actually being a separate instance/process of the same executable function. In some cases, complicated applications may have multiple processes; for example, Visual Studio runs a separate process when it compiles code from when it displays the IDE (design interface.) However, most often, a given application is running from a single process; for example, no matter how many Microsoft Word windows you have open, only a single instance of winword.exe is running.

An **application** is a program which you interact with on the desktop. This is what you spend almost all of your time using on the computer. Internet Explorer and Firefox web browsers, Microsoft Word, Excel, PowerPoint, iTunes, and Skype—they are all applications.

Sometimes, processes may run in the background without interacting with the desktop, but without being installed as a service. Many device drivers with enhanced features do this as they operate. For example, a touchpad driver will usually have a process that runs when a user signs in and handles the special features of the touchpad, but isn't a service and doesn't show any window to the user.

Sometimes an application may depend on a certain service. Printing from any program requires that the print spooler service be active. Installation packages (.msi installers) require that the Windows installer service be running.

Processes usually exit, or stop running and are unloaded from memory, when an application is closed. However, this is not always the case. Some programs such as a downloader or backup programs may continue to run in the background without displaying a visible window. Antivirus programs are also an example of this—in addition to using a service, many antivirus applications run a process silently in the background and only display a window to the user when action is required.

TASK MANAGER

For those curious about what applications, processes, and services are running in your PC computer, you can use the Task Manager to see a list of the software components in current use.

Right-click anywhere on the menu bar and select Start Task Manager. The first tab is the Applications tab that displays a list of programs that are running. If you haven't launched any applications, this list will be empty. If you started an application and minimized it, or it is hidden behind another application, it will appear on the list.

Next, click on the Processes tab. Here you will see all process programs that are running in support of the applications you may have open. This tab is of particular interest if you suspect your computer is running slowly for some reason. The CPU and Memory columns show how much of the CPU resource is being consumed by each of the processes.

As an exercise, go to your computer and open the Task Manager. I want you to observe which processes are in the list. Then, launch a web browser, such as Chrome. Now switch back to the Task Manager Processes tab window and see how many Chome.exe processes are running. On my PC, Chrome has about fifteen processes open. Close Chrome and the processes will go away. Now open Firefox (or any other web browser you may have). See how many processes Firefox has running. On my PC, Firefox uses four processes. Then do the same for Internet Explorer. It has only two processes. You might want to think twice before you use Google Chrome for web browsing. What are all those processes doing? They are most likely reporting back to Google the information you are searching for, so they can push ads to you. When I looked at my laptop's Task Manager, Chrome.exe has 12 processes running in the background even before I launch Chrome. I've since uninstalled Chrome from my computers.

Now click on the Services tab. There you will see the essential operating system programs. This list will display all the services programs that are running, mostly those which are part of the core operating system. The Description and Group columns are nice to review.

One of the benefits of looking at the Processes and Services tabs in the Task Manager is to become familiar with the programs that are usually running on your computer. Later if you suspect some program is hogging your computer, you can look at this list and identify it. If you see something running that you suspect should not be there, you can terminate execution of that process. But be careful—if you terminate an essential process such as explorer.exe, your desktop will go blank or your internet connection will stop. Don't worry—just reboot your computer if your screen goes blank.

Skype is or was one of those programs that continues to run in the background, as seen by Task Manager, even after you've closed it. I learned about this because I would later try to launch Skype, and nothing would happen on my screen. After a few tries to start Skype, I looked into Task Manager and saw that several copies were already running. Using Task Manager, I selected End Task, and once all instances of the program had been closed, Skype opened as expected. In most cases that will resolve the issue and your program will launch.

MALWARE AND CYBERSECURITY

Cybersecurity is probably the number one source of issues I encounter when a client calls with computer problems. These problems usually take the form of not being able to access the internet, the browser going to another website than the one you clicked on, or just the hard disk LED is constantly blinking, and the computer is slow to respond to keyboard and mouse commands.

The terms Trojan, worm, and virus are often used interchangeably, although technically they are not the same. Viruses, worms, and Trojan horses are all malicious programs that can cause damage to your computer, but there are differences among the three, and knowing those differences can help you to better protect your computer from their often damaging effects.

Viruses, Trojans, and Rootkits are things that get into your computer for one purpose, and one purpose only—to disrupt your life and steal your information. A Trojan is a type of malware that is often disguised as a legitimate software program. Trojans are written and distributed by cyberthieves and hackers trying to gain access to your system and any system they can infect. These Trojans get into your computer often by tricking you through some form of social engineering. Once activated, Trojans can enable a cybercriminal to spy on you, steal your sensitive data, and gain backdoor access to your system. These actions can include:

- Deleting data
- Blocking data
- Modifying data
- Copying data
- Disrupting the performance of computers or computer networks
- Browser hijacking

Browser hijacking is a form of unwanted software that modifies a web browser's settings without a user's permission, to inject unwanted advertising into the user's browser. A browser hijacker may replace the existing home page, error page, or search engine with its own. These are often used to force hits to a particular website, increasing its advertising revenue.

Some browser hijackers also contain spyware. For example, they could install a software keylogger to gather information such as banking and e-mail authentication details. As you type your secret password, it is being sent to a hacker to record. Some browser hijackers can also damage the registry on Windows systems, often causing permanent damage.

Often, browser hijacking can be easily reversed, while other instances may be difficult to reverse. Various software packages exist to prevent such modification.

Many browser hijacking programs are included in software bundles that the user did not choose to install. They come with a new computer or are presented as offers in the installer for another program, often included with no uninstall instructions, or documentation on what they do, and they are presented in a way that is designed to be confusing for the average user in order to trick you into installing unwanted extra software. Adobe Acrobat Reader is known to install things other than just Reader. Read the installation screens carefully and uncheck the boxes that will install the offers.

There are several methods that browser hijackers use to gain entry to an operating system. Email attachments, files downloaded through suspicious websites, and torrents are common tactics that browser hijackers use to gain access to your computer.

Trojans are generally not capable of replicating themselves and infecting other computers on your network. They just take over your computer. They are often capable of automatically hiding from malware removal software tools that are trying to detect and remove them from your PC.

Viruses are capable of spreading to other computers. Probably the number one culprit for why your PC is running slowly is quite simply an uninvited virus and Trojan program. These virus programs do all sorts of mean, evil, nasty things like:

- Keyloggers—they record your keystrokes and report them back to a server, often located in a foreign country. If you have a key logger running on you PC, you can say goodbye to keeping your banking username and password secure.
- Browser tracking cookies—every time you visit a website with a web browser, most websites leave tracking cookies behind. Cookies are useful for most websites you frequent like your online banking. You don't have to enter your username each time, only your password. However, there are some bad tracking cookies that are used by shopping websites and by Google. The history of where you have been shopping is sold to marketing companies, in an effort to push ads targeted toward your shopping preferences, based on what you've been looking at.

Essential programs like your antivirus need to be running in the background all the time. But programs such as Skype and Facebook notifications, as well as many other social media applications, do not need to be running all the time. That is my personal opinion. When I'm working on my PC I really don't need to see popup ads or hear beeps alerting me when my friends are coming and going online. I really don't need to be notified, or should I say distracted, the moment I receive an incoming email message.

When browsing a website, if a popup says, "This app wants to send notifications to your computer," I check no immediately.

CHOOSING AN ANTIVIRUS PROGRAM

There are several dozen antivirus detection, prevention, and cleaning tools on the market. Some are totally free, while some offer a free version with an upgrade to their pro version. The better ones are available as a paid subscription, with prices around $29 to $49 per year. Some antivirus programs come preinstalled by the computer manufacturer when you purchase a new computer, and some come with the Windows operating system. Windows 10, for example, comes with its own Windows Defender and initializes itself for your protection the very first time you turn on a new computer. Keep in mind: having a really good antivirus program to keep your data, programs, and computer safe is more important than anything else I've discussed in this book.

It is equally important to have good antivirus protection on all your other computing devices that you connect to your home or office network. This includes your mobile phones (Android and iPhone), tablet PC, notebook PC, netbook PC, and Kindle devices. Why?

Any malware such as a virus or Trojan is the biggest threat to your data safety. Ransomware can be devastating. Malware is capable of infecting not only your computer, but if your computer is a home network, can push itself onto all of your connected devices. If your laptop, mobile phone, or tablet is connected to your home or office network, there is a chance that these nasty programs could propagate to these other devices. Or, if your laptop picked up a virus when you were using it at the coffee shop, it could then infect your other computers when it connects to your Wi-Fi at home.

So I strongly recommend installing a quality antivirus as the first line of defense program immediately. You know that old adage "You get what you pay for"? I don't want to knock free programs like AVG or any one of a number of free antivirus programs, but if you are not paying for the yearly subscriptions, you probably are not getting the depth and breadth of protection needed to really be safe. Windows Defender that comes with Windows 10 is probably the only exception. But don't rely on just one antivirus program.

People who write malware programs spend a considerable amount of time and effort to reverse engineer the leading malware detection software in an effort to find ways to circumvent being detected. It's a kind of cat and mouse game between the malware programmers and antimalware software writers. It is important to always update your antispyware program each time you launch them.

Suggestion 11: *Update the anti-spyware detection definitions every time you run the program.*

A good antivirus program should protect you against all types of malware, including viruses, ransomware, Trojans, worms, and spyware. It does so in real time. That means the program is always checking incoming sources, and it checks and stops malware before it can install itself on your computer.

The malware protection programs are installed and load every time your computer boots up, and they are always running in the background, checking what is coming in, going out, and anything that changes on your computer. The old adage "More is better" does not apply to malware detection programs. There should only be one installed malware program on your PC. What do I mean by installed on your PC? It means loaded and running as a background task. You can still have, and should have, other antivirus programs sitting on your hard drive as an application program that can be started and then exit as needed. These serve as a second opinion to the primary antivirus program. Start it up and let it run. When the program finds stuff to clean out, say okay, then exit the program.

As an IT consultant, one of my corporate accounts stated they really didn't need malware protection on their servers. "We'll be fine," the CEO said. "We've never had a problem and we're too small of a business for anyone to want to hack into our system." Their accountant asked me if they were vulnerable as their competitor was recently put out of business as the result of a Ransomware attack. A year later someone in their office or warehouse accidentally clicked on an email attachment, and a Ransomware program installed on their accounting server. Within seconds literally everything on that server was encrypted and rendered useless. When I say everything I mean absolutely every program and all the business data. The Ransomware also encrypted everything on the USB drive connected to the server, which was their database backup. If it were not for that fact that I had transferred a backup of their entire accounting system to an off-site server a few days before the incident, they would not be in business today. Can I say more? A comprehensive antivirus program with ransomware detection, along with an off-site backup regimen, cannot be overstated. Not to mention the importance of data backed up in the cloud.

Are you looking for me to recommend an antivirus program for you? With so many on the market that are indeed good, it's hard to make a recommendation. However, over the years I have come to trust some over others. Personally, I use ESET products on all my personal and business computers. I then use SuperAntiSpyware as a second opinion and as a cookie and Windows temporary file remover. I recommend reading the current reviews that are available online.

VIRUS SCANNERS

While having a good antivirus program running in the background all the time is essential, so too is periodically running an antivirus scanner. No one antivirus program can detect all the viruses circulating around the Internet. It's important to have a few virus scanning programs ready to run and use them routinely.

An antivirus program runs in the background all the time, whereas a virus scanning program is launched only when you want to check your computer for things that might have slipped by the main antivirus program. Once it's done scanning it stops running, and lets the main antivirus program continue to do its job.

Some seemingly harmless things that get deposited in your computer browser are cookies. Cookies are little data files that get saved on your computer. They come in through the web browser from websites you visit, including Google search, Bing search, and Yahoo search. Cookies save what you have been searching for, and they are also used to track which websites you have been visiting, especially shopping sites like Amazon. You might have noticed ads popping up on your mobile device for things you were searching for on your PC. Google, Double Click, and others use these tracking cookies to make money on your personal searching history, and often change your personal preferences for their advertisers and sponsors.

Scanning programs will identify and remove these tracking cookies, making it more difficult for your personal information to be tracked and sold to others.

A list of some virus scanners is included in the Appendix.

CLEANING OUT TRACKING COOKIES, MALWARE, AND TROJANS

Websites that you visit use cookies to keep track of your visits, and the companies that set these cookies in place sell their tracking systems to many, many website customers. They use sophisticated algorithms to gather their tracking cookies from all the places you visit to compile data on where you are shopping and sites you are visiting, and then they use that information to sell advertising targeted to your interest. For example, you use Google search to look for patio furniture, and then click on that website to see what they are offering. Cookies are left behind in your browser. You later open your browser and see read your news feed and see patio furniture ads on the right-hand side of the screen, often persisting for weeks. The magic starts with tracking cookies.

The use of tracking cookies has become so sophisticated that it may not be able to completely stop or prevent them from following your path. Cookies are an inevitable consequence of advanced technology applied to the art of selling. However, there are things you can do to slow them down.

Each web browser stores cookies in its own directory on your hard drive. Using the Settings icon on most browsers, you can clear out the browsing history (the sites you've visited) and their cookies. Firefox, for example, gives you the option to clear cookies automatically when you exit.

The browser war continues, and not all browsers adhere strictly to the JavaScript specifications (JavaScript is the programming language used inside a web browser). You might discover that some feature of a website doesn't work with a particular web browser. Computer users often have different browsers on their PC for visiting different websites, such as your banking portal. Clearing cookies from each web browser

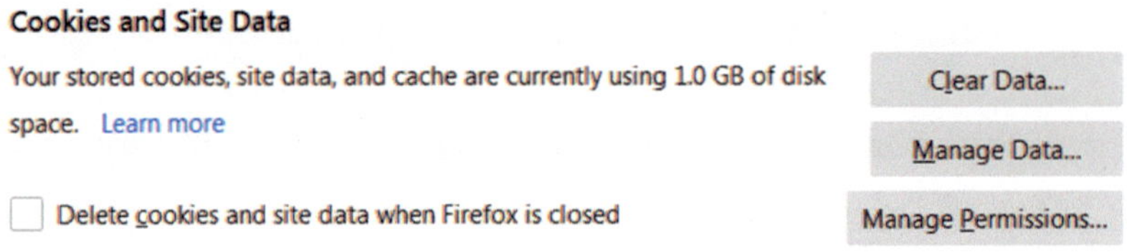

Figure 24: Firefox Cookies and Site Data Options

can be done from within each web browser. You can also clear cookies from any web browser using a scanning program that cleans all cookies. A program I like to use is the free version of SuperAntiSpyware.

CLEANING UP OPERATING SYSTEM STUFF

Most new Windows-based computers come preloaded with a whole bunch of extras or stuff you probably don't really need or want. The forces of marketing and advertising are at play, even before you unbox your brand new computer and turn it on for the first time. Many computer manufacturers get paid to advertise for third party companies (a third party means another business), trying to sell or upgrade your software to their brand or version. This is done with popups that suggest their product, or offer you discounts if you click now to purchase. If you run a virus scanner such as SuperAntiSpyWare on a fresh, never used Windows 10 PC, it will trap a dozen or so cookies ready to start tracking you. Don't be surprised, because Microsoft has started the process of reselling your personal information from the very first time you turn on a new computer.

The most common of these preinstalled bonus programs is virus protection software. Windows comes with its own basic antivirus protection called Windows Defender. Companies like Norton, McAfee, and Kaspersky often include the installation of their antivirus programs in a new PC without asking you. The program is offered as a free trial with your PC purchase, for a limited time. One of the tasks on my to-do list after configuring the settings and preferences of a new computer is to uninstall all the free programs and games. Instructions for uninstalling these programs can be found on page 70, .

As the complexity of software has increased, so too have the myriad of options, checkboxes, and settings that are possible. Often a simple checkbox might take hours to locate. My best recommendation, one that separates the novice from the guru, is persistence. Don't give up looking and trying the various options. If you are not certain what it will do, try it out and learn for yourself. You can always change the setting back the way it was.

If you think you are spending too much time hunting down a "How can I do that?" question, don't forget you can always ask the internet. A quick search on Google is often the quickest way to remember how to do something.

CLEANING UP APPLICATION STUFF

As you use your computer over time, application programs that you run, or programs that you tried out but did not use, might begin to take up space on the hard drive. This in itself is not an issue unless your hard drive is already getting full. Many programs that you installed will also store temporary data in hidden directories and the system registry, and this data hangs around on your computer for years, serving no purpose.

From time to time I like to clean up my hard disk of unused programs and data. Windows has a tool provided for just this purpose. It's called Disk Cleanup as seen in Figure 25. You can access Disk Cleanup by searching for Disk Cleanup in the search bar. It looks like this:

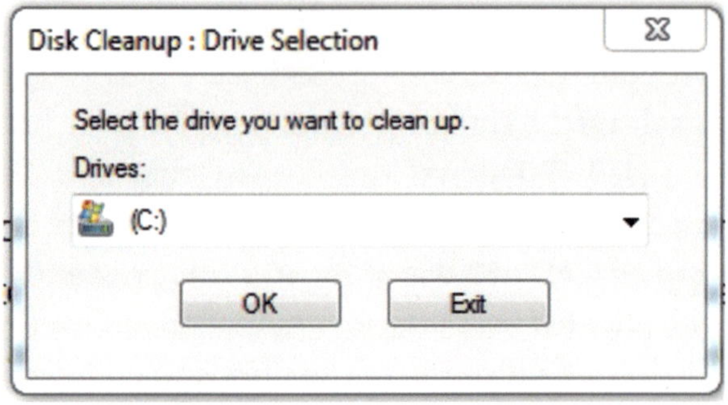

Figure 25: Disk Cleanup Menu

Your default hard disk is already selected, and you can also clean up other hard drives on your computer such as a USB backup drive or USB memory stick. However, I don't recommend cleaning up these devices as they probably don't need to be cleaned. The windows operating system keeps track of programs you use and don't use each time.

Press OK to continue and Disk Cleanup will determine which files will be cleaned. Once it's done, you will see a list of directories and file types to be deleted. See Figure 26.

Scroll through the list and check each file type to delete. I typically delete everything including the Recycle Bin. Once you've checked off all the files to delete, press OK, and confirm that you want to delete these files. If you don't trust the Disk Cleanup program, you can View Files beforehand to see the directory and the files that will be deleted, but to be honest, it's not necessary.

Another area often neglected for a spring cleaning is to clean up the system files that Windows saves when performing tasks such as

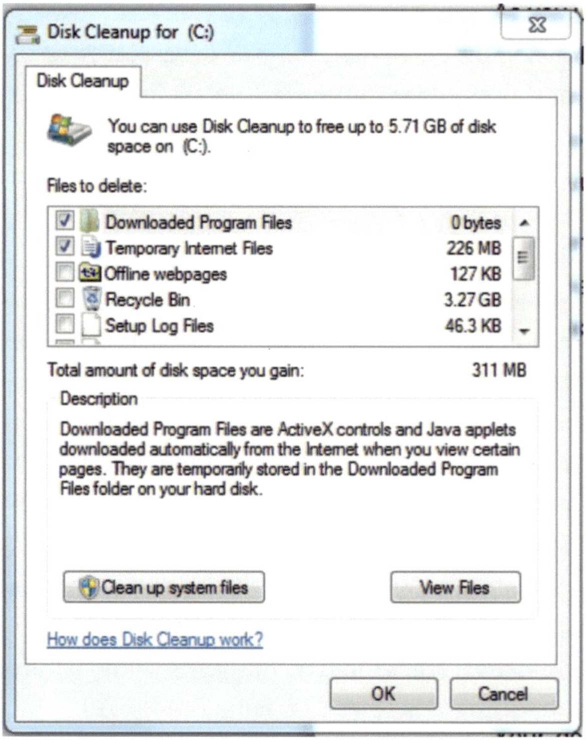

Figure 26: Select File Types to Clean

Windows Updates. While it is safe to delete everything in the system files list, it is recommended that you DO NOT delete the Windows ESD Installation Files. These ESD files are used for resetting your PC to its factory default settings. If you delete these files, you'll have more disk space—but you won't have the files necessary to reset your PC. Should you need to, you may download Windows 10 installation media if you ever want to reset it.

It is recommended not deleting this, unless you really need the few gigabytes in hard disk space. Deleting this will make your life harder if you ever want to use the *reset your PC* feature in the future.

REMOVING OR UNINSTALLING A PROGRAM

There are right and wrong ways to remove or delete programs that you no longer use. Removing a program frees up space on your storage device. Remember, having programs on your hard drive does not slow down your computer. As long as you have plenty of storage space available, it isn't necessary to remove an application. I have programs installed that I've long forgotten what they were used for. I often install free or trial programs just to see what they are, and when I'm done, I remove them from my computer. It's just part of my routine to keep good habits.

The cleanest way to uninstall an application program you no longer want on your computer is use a built-in tool that deletes the programs properly. It also cleans up the system registry where information about the program is saved and associated user data. Having the uninstall process clean the registry can be important in the long run as it too can become quite a large file, filled with information and references to programs long gone.

To safely uninstall a program, search for Add or Remove Programs in the case of Windows 10, or Uninstall a Program for the older Windows 7—which you should no longer be using, because Microsoft stopped supporting Windows 7 on January 14, 2020.

When the Control Panel -> Programs -> Program Features screen appears, you will see a list of installed application programs. The list will tell you the program name, the manufacturer of the program, the date you installed it, and the version.

It is useful to observe the installation date. Why? One of the ways malware is installed is actually through a simple installation process, which is recorded on this list. If you see a bunch of games, utilities, or other programs that you don't recognize or remember having installed yourself, looking at the installation date could provide you with the clue that it was not installed by you, but by a rogue installer program.

It could happen that when you purchase an application, or download one from the internet, that it will install other programs for you. Adobe Acrobat has done that in the past along with AVG antivirus and other partner programs. If you didn't see the checkbox to not install a companion program, voila, you now have a new web browser, or something else you didn't really want installed.

To uninstall the unwanted program, simply click on the program name and the Uninstall/Change button will activate at the top of the screen. Press Uninstall and you will be asked "Are you sure you want to uninstall this program?" Press OK to continue, and the program will be properly uninstalled.

Problems can arise when simply deleting a program by moving the program folder to the trash can. No. No. No. Don't do that. That is not how to remove programs.

DEFRAGMENTING YOUR HARD DRIVE

As data is read, deleted, and written to and from the hard drive, a disk controller program built into your hard drive manages these processes. When programs are deleted, they free up space which is then reused the next time the hard drive writes data to the rotating magnetic media or solid-state drive. However, if the amount of storage needed is not contiguous, the data will be written in segments on the storage device. Later, when your computer wants to read this data or program into memory, the controller has to work a bit harder to assemble all the fragments of the information back into a contiguous file. This can slow down the rate at which programs and data is transferred. In older, slower computers, this slowdown was very notable.

In operating systems from Windows XP and prior versions, it was recommended to manually run the disk defragment program. In fact, it was kind of fun to watch the fragments of information be reassembled into contiguous files as the graphic interface was colorful and dynamically updating as the program ran. Figure 27 below is a screenshot of what this display looked like on the screen. Red meant defragmented data, while blue was contiguous data. As you watched, you could see the blue areas growing as data was moved and relocated.

Starting with Windows 7, the defragment function runs automatically for you on a schedule, usually once per week. However, it might be a good idea to double check.

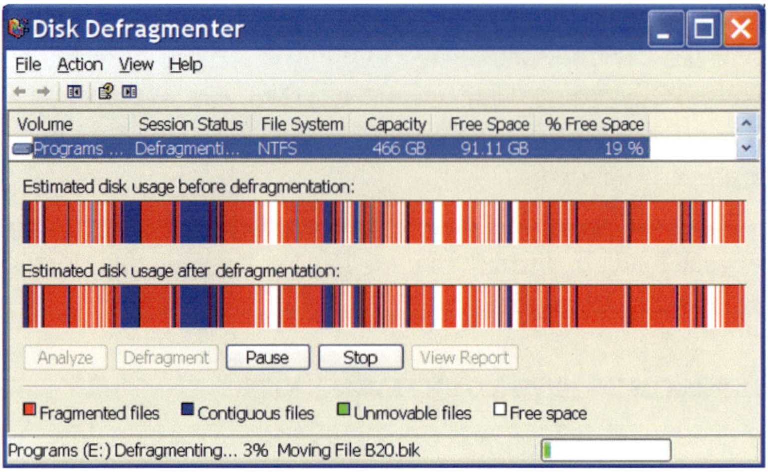

Figure 27: Disk Defragmenter Moving Files View

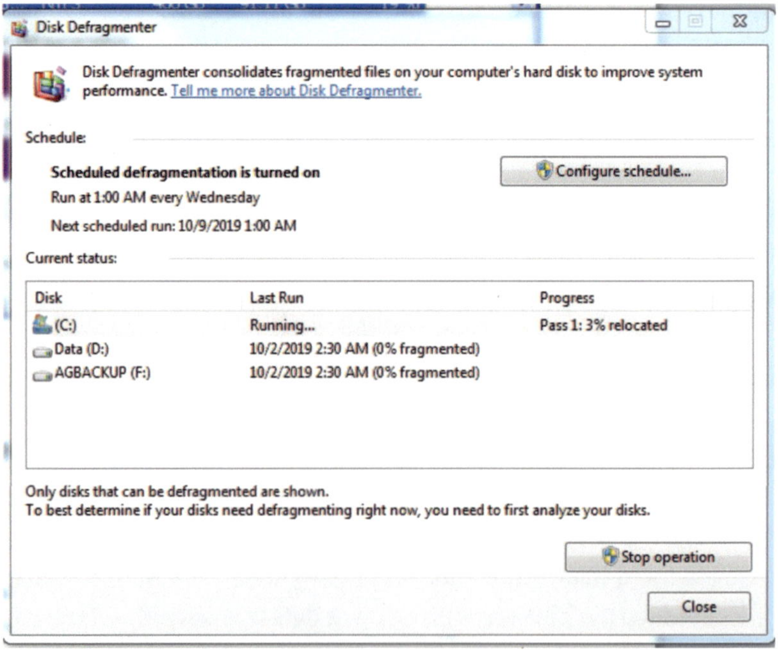

Figure 28: Disk Defragmenter Schedule View

Computers with a solid-state drive (SSD) should not be defragmented. Doing so can cause unnecessary wear that might reduce its lifespan. SSDs are able to read blocks of data that are spread out over the drive just as fast as they can read those blocks that are adjacent to one another. My C drive was NOT appearing on the list of drives to be defragmented—I forgot that it was an SSD, but the operating system knew that it was an SSD and did not enable it.

To launch the disk defragment program, search for, you guessed it, Disk Defragmenter. It looks like this:

RUNNING WINDOWS UPDATES

Together with keeping the hard disk clean of unwanted or unneeded information and programs, downloading and installing the Windows updates is a very important good habit. Hackers spend an inordinate amount of time and effort to find vulnerabilities in the Windows operating system. They also look for vulnerabilities in antivirus programs, and in just about any other application program you might have on your computer. For a long time, Adobe Flash and Acrobat Reader PDF files were a very vulnerable vector for getting malware into your computer.

As these vulnerabilities and exploits are uncovered by the vendors, they issue updates. The Windows operating system update is your first line of defense. Windows updates usually run automatically, and you might want to confirm that it is set to update automatically.

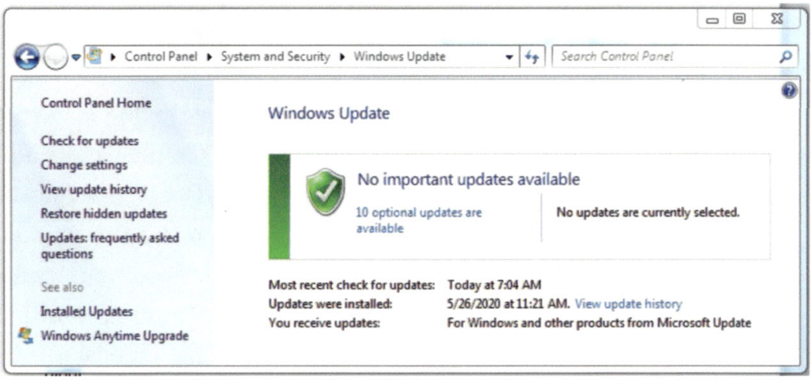

Figure 29: Windows Update Control Panel

To configure Windows Update, navigate to the Control Panel, System and Security, then Windows Update screen as shown in Figure 29. In the Change settings page, select the preferred method and times to run the update checker. I have mine set for Install Updates Automatically and to do so every day at 3:00 am. This is another good reason why you should leave your PC turned on all the time—the operating system does cleanup and maintenance overnight when you are not using your computer.

Windows Update will update any of your other Microsoft programs that you might have installed such as Word, Excel, and PowerPoint. It's a good idea to look at the settings for your other applications and make sure you are getting those vendors updates. Skype, for example, will automatically check and install program updates. QuickBooks accounting as well runs updates automatically. In order to ensure your PC is protected, always look for and install your application updates.

WINDOWS EXPLORER

I would like to make mention of Windows Explorer—the software tool used to visually navigate the system file structure. It is used to see the files stored on the hard drive and external drives such as a USB thumb drive or memory stick. Windows Explorer provides a lot of useful information accessible from Properties. Select a folder or a file and using the right-mouse, click Properties.

The folder structure displays information in a tree-like structure. But before you start using Windows Explorer, I would like you to change an option setting that will allow you to see the file extension information along with the filename information. The file extension is the three letters following the filename such as myDocument.docx. It is very useful to be able to see the extension when you are working with a file. For example, I may have a word document called myDocument.docx. If I create a PDF from myDocument, it will be called myDocument.pdf. Unless you can see the file extension, you won't know which filename to select, if you want to send it as an email attachment.

Depending on the version of operating system you have, you can set this option checkbox either from the Explorer Organize dropdown, selecting Folder and Search options and then clicking on the View tab. There look for Display the full path in the title bar and turn off Hide extensions for known file types. If your File explorer has the ribbon bar on the top, select View and then in the Show/Hide section, check the box for File name extensions.

In Summary

I had a calculus professor in college who repeatedly asked the students to please stay awake and listen to what he was talking about. He would say, "Today you may not have any idea what I'm talking about. But one day in the future, while you are stopping to tie your shoe, you'll remember something I explained, and the light will go on. Aha—that's what he meant." I know because it happened to me.

You have learned more about your personal computer than you probably thought you would want to know. Some things may make sense today. Some things you'll have no idea how to apply to your daily skill set. Maybe, one day in the future, you will be reading something about computers, or a friend will make a suggestion to you and you'll say, "Aha—I read about that someplace. Now I know what it means and why it might be important and useful."

You can now sit down in front of your "friend" who is going to be helping you with all kinds of urgent, important, and vital activities. Smile and say, "Hello dear computer. I look forward to working with you today." Having a positive attitude toward your computer will have a positive effect on how your computer responds to you. All things are connected to one another and we are connected to all things. And so it is.

Summary of Suggestions

Glossary of Terms

Chassis	A computer case, tower, or cabinet, it is the enclosure that contains the components of a personal computer
Application Software	Application software (app for short) is a program or group of programs designed for end users. Examples of an application include a word processor, a spreadsheet, an accounting application, a web browser, an email client, a media player, a file viewer, simulators, a console game, a photo editor or a UO NARCIS. The collective noun application software refers to all applications collectively. This contrasts with system software, which is mainly involved with running the computer.
Backup	A duplicate copy of a file, program, folder, or disk used if the original is damaged, corrupt or lost.
Backup Media	Can be another computer, floppy disk(s), CD(s), DVD(s) or any other recordable media for computers.
BIOS	BIOS is an acronym for basic input/output system and is also known as the System BIOS, ROM BIOS or PC BIOS is firmware used to perform hardware initialization during the booting process (power -on startup), and to provide runtime

services for operating systems and programs. The BIOS firmware comes preinstalled on a personal computer's system board, and it is the first software to run when powered on. The name originates from the basic input/uutput system used in the CP/M operating system in 1975. The BIOS originally proprietary to the IBM PC has been reverse engineered by companies looking to create compatible systems. The interface of that original system serves as a de *facto* standard.

Blackout	A power outage, also called a power cut, a power out, a power blackout, power failure or a blackout, it is the loss of the electrical power network supply to an end user.
Booting	In computing, booting is the process of starting a computer. It can be initiated by hardware such as a button press, or by a software command. After it is switched on, a computer's central processing unit (CPU) initially has no software in its memory, so some process must load software into memory before it can be executed. This may be done by hardware or firmware in the CPU, or by a separate processor in the computer system.
Brownout	A drop or reduction in voltage in an electrical power supply, so named because it typically causes lights to dim
CPU	A central processing unit (CPU), also called a central processor or main processor, is the electronic circuitry within a computer that executes instructions that make up a computer program. The CPU performs basic arithmetic, logic, controlling, and input/output (I/O) operations specified by the instructions in the program.

Full Backup	Backing up everything on your computer. This includes data, software and operating system.
Hard Disk Drive	A hard disk drive (HDD), hard disk, hard drive, or fixed disk is an electro-mechanical data storage device that uses magnetic storage to store and retrieve digital data using one or more rigid rapidly rotating platters coated with magnetic material. The platters are paired with magnetic heads, usually arranged on a moving actuator arm, which read and write data to the platter surfaces. Data is accessed in a random-access manner, meaning that individual blocks of data can be stored and retrieved in any order. HDDs are a type of non-volatile storage, retaining stored data even when powered off.
I/O	Input/output or I/O is the communication between an information processing system, such as a computer, and the outside world, possibly a human or another information processing system. Inputs are the signals or data received by the system and outputs are the signals or data sent from it. The term can also be used as part of an action; to "perform I/O" is to perform an input or output operation.
Incremental Backup	A backup of only what has changed since the last backup.
Integrated Circuit	An integrated circuit or monolithic integrated circuit (also referred to as an IC, a chip, or a microchip) is a set of electronic circuits on one small flat piece (or chip) of semiconductor material that is normally silicon. The integration of large numbers of tiny MOS transistors into a small chip results in circuits that are orders of magnitude smaller, faster, and less expensive than those con-

	structed of discrete electronic components. The IC's mass production capability, reliability, and building-block approach to circuit design has ensured the rapid adoption of standardized ICs in place of designs using discrete transistors. ICs are now used in virtually all electronic equipment and have revolutionized the world of electronics. Computers, mobile phones, and other digital home appliances are now inextricable parts of the structure of modern societies, made possible by the small size and low cost of ICs.
Malware	Malware (short for malicious software) is any software intentionally designed to cause damage to a computer, server, client, or computer network. By contrast, software that causes unintentional harm due to some deficiency is typically described as a software bug. A wide variety of types of malware exist, including computer viruses, worms, Trojan horses, ransomware, spyware, adware, rogue software, and scareware.
Motherboard	A motherboard (also known as the mainboard, main circuit board or system board) is the main printed circuit board (PCB) found in general purpose computers and other expandable systems. It contains and connects the many of the crucial electronic components of a system, such as the central processing unit (CPU) and memory, and provides connectors for other peripherals.
Operating System	An operating system (OS) is system software that manages computer hardware, software resources, and provides common services for computer programs.

RAM	Random-access memory is a form of computer memory that can be read and changed in any order, typically used to store working data and machine code. A random-access memory device allows data items to be read or written in almost the same amount of time irrespective of the physical location of data inside the memory.
Restore	To copy the backed-up version of information back to its original location.
Solid State Drive	A solid-state drive (SSD) is a solid-state storage device that uses integrated circuit assemblies to store data persistently, typically using flash memory, and functioning as secondary storage in the hierarchy of computer storage. It is also sometimes called a solid-state device or a solid-state disk, even though SSDs lack the physical spinning disks and movable read–write heads used in hard drives (HDD) or floppy disks.
Switching power supply	An electronic power supply that incorporates a switching regulator to convert electrical power efficiently.
Task Manager	Task Manager, previously known as Windows Task Manager, is a system monitor and startup manager that is included with Microsoft Windows systems. It provides information about computer performance and running software, including name of running processes, CPU load, commit charge, I/O details, logged-in users, and Windows services. Task Manager can also be used to set process priorities, processor affinity, and start and stop services, and forcibly terminate processes.

Tracking Cookie	An HTTP cookie (also called web cookie, Internet cookie, browser cookie, or simply cookie) is a small piece of data sent from a website and stored on the user's computer by the user's web browser while the user is browsing. Cookies were designed to be a reliable mechanism for websites to remember stateful information (such as items added in the shopping cart in an online store) or to record the user's browsing activity (including clicking particular buttons, logging in, or recording which pages were visited in the past). They can also be used to remember arbitrary pieces of information that the user previously entered into form fields such as names, addresses, passwords, and credit-card numbers.
Trojan	In computing, a Trojan horse, or trojan, is any malware which misleads users of its true intent. The term is derived from the Ancient Greek story of the deceptive Trojan Horse that led to the fall of the city of Troy.
UPS	An uninterruptible power supply or uninterruptible power source (UPS) is an electrical apparatus that provides emergency power to a load when the input power source or mains power fails. A UPS differs from an auxiliary or emergency power system or standby generator in that it will provide near-instantaneous protection from input power interruptions, by supplying energy stored in batteries. The on-battery run-time of most uninterruptible power sources is relatively short (only a few minutes) but sufficient to start a standby power source or properly shut down the protected equipment. It is a type of continual power system.

USB — Universal Serial Bus (USB) is an industry standard that establishes specifications for cables and connectors and protocols for connection, communication and power supply (interfacing) between computers, peripherals and other computers.

- 3 1/2" floppy can hold 1.44mb
- CDs can hold 650-800mb
- DVDs can hold 4.7gb
- External hard drive can hold various amounts of information. How much will depend on the size of drive you have

References and Resources

Over the years, countless authors, technology writers, editors, and individuals have contributed to the wealth of information available for you to find on the internet. YouTube has hundreds of how-to videos on just about every aspect of the personal computer, covering hardware, operating systems, application programs, and explaining how to use them. Searching the internet for useful information should be your go-to resource to learn how things work, and serves as a reference or reminder. One caveat—read more than one online resource for the answer to your question. Another caveat—question everything you read on the internet. Always consider the source.

ANTIVIRUS SCANNING PROGRAMS

It is not possible to list all the free and low-cost antivirus scanning programs available. There are too many and they come and go frequently. Some are free but when they find things that should be removed, you have to pay for an upgrade. Read carefully before you download and install any program. Often the text will read *free download*, but then there is a subscription fee to actually use the program. Before you buy, do your research and read the reviews. I've listed only a few programs that I use for malware scanning:

SuperAntiSpyware—Download free edition
https://www.superantispyware.com

Malware Bytes—Get the free version
https://www.malwarebytes.com/

Trend Micro "Housecall" free edition
https://www.trendmicro.com/en_us/forHome/products/housecall.html

OTHER RESOURCES

What Is a CPU Core? A Basic Definition
https://www.tomshardware.com/news/cpu-core-definition,37658.html

Block Diagram of a Motherboard. Shown on page 14
http://itexpertinbd.blogspot.com/2013/07/laptop-motherboard-block-diagram.html

Error Correcting Memory
https://www.techopedia.com/definition/2774/error-correcting-code-memory-ecc-memory

How to Fix the Blue Screen of Death
https://recoverit.wondershare.com/computer-problems/fix-blue-screen.html

General articles on security and how to be safe using the Internet
https://dataprot.net/

Glossary of Terms is mostly drawn from https://en.wikipedia.org

Stock Photos by Pexels
https://www.pexels.com/

Cover Photo by iStockPhoto
https://www.istockphoto.com/

Photography by Avram Grossman, Author

Acknowledgements

Thank you to all my former and current customers who have provided me with the opportunity to be exposed to a wide breadth of computer and technical issues along with the variety of computer user personalities. These experiences enabled me to observe the patterns in how and why technical stuff failed to work, and how to quickly diagnose and repair those issues so you can get back to work.

I would like to thank Dr. Jim Turrell, Minister of the Center for Spiritual Living in Costa Mesa, California. He is a spiritual leader extraordinaire and a book writing mentor and coach. His encouragement and positive "You can do it!" mindset provided me with the encouragement and belief in myself that encouraged me to write this book. "And so it is."

I would like to thank Alan Garner, a well-known author in his own right. I first met Alan at a book signing on Balboa Island, Newport Beach, thirty-five years ago. His publication entitled *Everything Men Know about Women* was a tremendous success. It was a blank book. He has since written other books with words printed on the pages that demonstrate his skills and humor. I thank you Alan for your encouragement and guidance and suggestions in choosing a title for this book.

I want to thank Kit Brown. To my recent surprise, she has been a friend of Alan Garner's for about as long as I've known Alan. She never knew I knew Alan. It's a small world. Kit is a like-minded computer tinkerer, software engineer, armchair physicist, and helped review this work and provided me with some great suggestions for enhancing its content.

About the Author

Avram Grossman started E-Novative, Inc., in 1999, as a website design and development business, writing custom internet-based websites and database programs for e-commerce clients. As the internet environment changed, E-Novative too continued to evolve. Today, Mr. Grossman provides IT technical solutions for complex business models that want to incorporate the myriad of third-party technology platforms that "integrate technology for your business, the way you do business."

Prior to forming E-Novative, Mr. Grossman was the director for technology business development at Samsung America where he started one of the first internet ecommerce development services in North America, along with helping to pave the way to allow the use of credit cards for online purchases. Along with starting SAILAhead Internet Services for Samsung, Mr. Grossman provided a technical perspective and analysis of new and emerging technology business investments and assisted the R&D team in Korea with new product development.

As the Director of Research and Development at Toshiba America, Mr. Grossman led a team that developed cutting-edge technology solutions for tomorrow using today's technology. He created one of the first touch-panel pen notebook PCs; developed the application of spread-spectrum radio for what is now known as the 802-11 or Wi-Fi standard; and established a patent and technology outreach program bridging North American inventors with product developers in Japan.

As an entrepreneur, Mr. Grossman invented the CardGrabber Business Card Scanning System, developed a biomedical device for high-risk home fetal monitoring, and is co-inventor of the One-For-All Universal Remote Controller. He has a number of other patents to his name.

As European Technical Support for Itel International based in Frankfurt, Germany, Mr. Grossman was one of the few people in the world who knew the AS/5 (an IBM 370/158) mainframe operating console inside and out. He was the final tech support person in the chain when there was a critical mainframe computer problem. If he couldn't find and fix the problem, there was nobody else he could call for help. His response to every problem was: "Failure is not an option."

Mr. Grossman provides IT consulting services, resolving complex problems and developing complex software solutions.

HOW TO CONTACT THE AUTHOR

You may contact the author through E-Novative, Inc. at http://www.e-novative.com

Made in the USA
Monee, IL
05 November 2026

67387501R00064